The Baseball Mysteries

The Baseball Mysteries: Challenging Puzzles for Logical Detectives is a book of base-ball puzzles, logical baseball puzzles. To jump in, all you need is logic and a casual fan's knowledge of the game. The puzzles are solved by reasoning from the rules of the game and a few facts.

The logic in the puzzles is like legal reasoning. A solution must argue from evidence (the facts) and law (the rules). Unlike legal arguments, however, a solution must reach an unassailable conclusion.

There are many puzzle books. But there's nothing remotely like this book. The puzzles here, while rigorously deductive, are firmly attached to actual events, to struggles that are reported in the papers every day.

The puzzles offer a unique and scintillating connection between abstract logic and gritty reality.

Actually, this book offers the reader an unlimited number of puzzles. Once you've solved a few of the challenges here, every boxscore you see in the papers or online is a new puzzle! It can be anywhere from simple, to complex, to impossible.

- For anyone who enjoys logical puzzles.
- For anyone interested in legal reasoning.
- For anyone who loves the game of baseball.

Jerry Butters has a BA in mathematics from Oberlin College, and an MS in mathematics and a PhD in economics from the University of Chicago. He taught mathematics for two years at Mindanao State University in the Philippines as a Peace Corps volunteer. He taught economics for five years at Princeton University. For most of his career, he worked on consumer protection cases and policy issues at the Federal Trade Commission. In his retirement, he has become a piano teacher and performer. He enjoys hobbies ranging from reading Chinese to practicing Taiji. This book is an outgrowth of another of his hobbies—his love of designing and solving puzzles of all sorts.

Jim Henle has a BA in mathematics from Dartmouth College and a PhD from M.I.T. He taught for two years at U. P. Baguio in the Philippines as a Peace Corps volunteer, two years at a middle school as alternative service, and 42 years at Smith College. His research is primarily in logic and set theory, with additional papers in geometry, graph theory, number theory, games, economics, and music. He edited columns for *The Mathematical Intelligencer*. He authored or co-authored five books. His most recent book, *The Proof and the Pudding*, compares mathematics and gastronomy. He has collaborated with Jerry on puzzle papers and chamber music concerts.

AK Peters/CRC Recreational Mathematics Series

Series Editors

Robert Fathauer
Snezana Lawrence
Jun Mitani
Colm Mulcahy
Peter Winkler
Carolyn Yackel

Mathematical Puzzles
Peter Winkler

X Marks the Spot
The Lost Inheritance of Mathematics
Richard Garfinkle, David Garfinkle

Luck, Logic, and White Lies
The Mathematics of Games, Second Edition
Jörg Bewersdorff

Mathematics of The Big Four Casino Table Games
Blackjack, Baccarat, Craps, & Roulette
Mark Bollman

Star Origami
The Starrygami™ Galaxy of Modular Origami Stars, Rings and Wreaths
Tung Ken Lam

Mathematical Recreations from the Tournament of the Towns
Andy Liu, Peter Taylor

The Baseball Mysteries
Challenging Puzzles for Logical Detectives
Jerry Butters, Jim Henle

Mathematical Conundrums
Barry R. Clarke

Lateral Solutions to Mathematical Problems
Des MacHale

Basic Gambling Mathematics
The Numbers Behind the Neon, Second Edition
Mark Bollman

For more information about this series please visit: https://www.routledge.com/AK-PetersCRC-Recreational-Mathematics-Series/book-series/RECMATH?pd=published,forth coming&pg=2&pp=12&so=pub&view=list

The Baseball Mysteries
Challenging Puzzles for
Logical Detectives

Jerry Butters
Jim Henle

CRC Press
Taylor & Francis Group
Boca Raton London New York

CRC Press is an imprint of the
Taylor & Francis Group, an **informa** business

AN A K PETERS BOOK

First edition published 2023
by CRC Press
6000 Broken Sound Parkway NW, Suite 300, Boca Raton, FL 33487-2742

and by CRC Press
4 Park Square, Milton Park, Abingdon, Oxon, OX14 4RN

CRC Press is an imprint of Taylor & Francis Group, LLC

© 2023 Jerry Butters, Jim Henle

ISBN: 978-1-032-36548-0 (hbk)
ISBN: 978-1-032-36505-3 (pbk)
ISBN: 978-1-003-33260-2 (ebk)

DOI: 10.1201/9781003332602

Publisher's note: This book has been prepared from camera-ready copy provided by the authors.

Contents

Visit the website:
www.science.smith.edu/~jhenle/baseball
for comments, clarifications, and additional
puzzles from the authors and readers

Send us yours at
gerardrbutters@gmail.com
jimhenle@gmail.com

and for a glossary of baseball terms.

Preface

This is a book of puzzles, logical puzzles. They're all about base-ball. They're strictly logical. They're never about what should happen, what probably happened, or what could have happened. They're al-ways about

what *must* have happened.

To enjoy the puzzles, you don't need to know any baseball history, or strategy, or statistics. A casual knowledge of the game is all you need to get started—

> 9 men on a team,
> 3 outs in an inning,
> the meaning of runs, hits, and runs batted in,
> walks, strikeouts, stolen bases, ...

When a special rule is needed, we introduce and explain it. There is an extensive glossary for all terms, ordinary and special.

Here's an example of the "casual knowledge" that's useful:
If there's a player on second base[1]

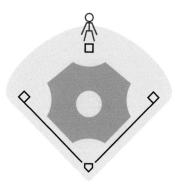

---

[1]The illustrations here are from the authors' "Baseball Retrograde Redux," *The Mathematical Intelligencer*, 44(4), 2022

and the batter hits a triple,

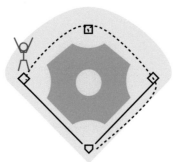

then it's clear that the player who was on second must have either scored a run or made an out. That's logically true because under baseball rules one runner can't pass another runner on the base paths.

But knowing a lot about baseball can actually get in the way. Suppose, for example, there's a man on third base,

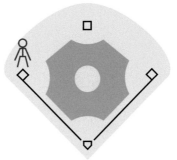

and the next player up hits a double. Knowing baseball, you expect that the player on third will score. But that's not a strictly *logical* conclusion. It's not against the rules for the man on third to just stay there!

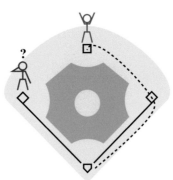

Baseball, for puzzling, is special. It has a clear structure—

> players run around bases,
> never more than one on a base,
> scoring only when they reach home plate, . . .

and each game is reported with a rich mix of data—

> home runs, stolen bases, errors, . . .
> innings pitched, strikeouts, at-bats, . . .
> number of players left on base, . . .

Put together, these can make it possible to reconstruct specific details of the game. Lovely, intricate puzzles can be constructed in a way that is inconceivable in any other sport.

We hope you enjoy them!

Jim and Jerry

Acknowledgments

We owe much, first of all, to the late Raymond Smullyan, whose wonderful books on chess retrograde analysis, *The Chess Mysteries of Sherlock Holmes*[1] and *The Chess Mysteries of the Arabian Nights*[2], opened our eyes to possibilities of logical problems in the context of a game with complex rules.

We are also indebted to the editors of the AK Peters/CRC Recreational Mathematics Series for continuing the vision of Alice and Klaus Peters and for their confidence in us. We are especially grateful for the assistance of Mansi Kabra in the smooth transition of manuscript to book.

We thank Springer-Nature for allowing puzzles from our columns in *The Mathematical Intelligencer* to appear in this volume.

We have been supported by many in our efforts. In the case of our wives, Ettie B. and Portia H., by their calming presence and their inexplicable tolerance. Special thanks goes to Sam Perkins who read one entire version and left us with extensive notes.

3

[1] Alfred A. Knopf, New York, 1979
[2] Alfred A. Knopf, New York, 1981
[3] Photo by Ettie.

CHAPTER 1

The Game Begins

1.1 IN PLAIN SIGHT

Jerry: Jim, we've been great friends for more than fifty years, and
we know each other very well, but I have a hobby that I don't
think I've ever told you about. Every day in the baseball
season, I like to study the box scores of the games. I like to
see how my favorite teams and players are doing, but that's
not what my hobby is about. Instead, I study the box scores
to look for the hidden information they contain.

Jim: Wow. Fifty years.

What do you mean, "hidden information?"

Jerry: Well, the box score only presents aggregate information
about the game. For example, it tells us for each player how
many at-bats, hits, runs, and runs batted in he got in the
game. But it doesn't say in which at-bat or which inning he
got his hits and scored his runs. The box score might tell us
that the visiting team rallied to tie the game in the top of the
ninth, but it doesn't tell us who scored in the ninth and who
drove in the tying run. To be precise, it doesn't tell us
explicitly. But often we can deduce this more granular
information. Often we can recover some of the specific
events in the game that were used to create the box score.

DOI: 10.1201/9781003332602-1

Jim: Whoa! That sounds really interesting! I never thought of that before.

But ... what can you deduce? Can you give me an example?

Jerry: Consider this box score, which I made up:[1]

Cats   Dogs

	ab	r	h	bi		ab	r	h	bi
Abe ss	5	0	0	0	Amsler 1b	3	0	1	0
Brown 3b	4	0	1	0	Blanco cf	3	0	0	0
Castro cf	4	1	1	0	Clark lf	4	0	1	0
Dominguez c	4	1	2	1	Durand rf	3	0	1	0
Engel rf	4	0	1	1	Emerson 2b	2	0	0	0
Frank 1b	4	0	0	0	Foster 3b	3	0	1	0
Guzman lf	3	0	2	0	Gato c	3	1	2	1
Hall 2b	4	0	0	0	Herrera ss	2	0	0	0
Iglesias p	4	0	0	0	Ito p	3	0	1	0
Totals	36	2	7	2	Totals	26	1	7	1

```
Cats  ··  1 0 0   0 0 0   0 1 0--2
Dogs  ··  0 1 0   0 0 0   0 0 0--1
```

	ip	h	r	er	bb	so
Iglesias (W)	9	7	1	1	4	4
Ito (L)	9	7	2	2	1	3

```
LOB-Cats 8, Dogs 2, HR-Gato, 3b-Guzman, Durand
2b-Amsler, Foster, Guzman, Dominguez(2), SB-Emerson,
DP-Dogs 2
```

Jerry: As you can see, the Cats scored runs in the first and eighth innings. The box score lists the batters according to the order in which they hit. We can see that Castro and Dominguez each scored a run, and Dominguez and Engel each batted in a run. But suppose we want to know who scored the winning run in the eighth, and who drove it in. See if you can figure that out.

---

[1] For readers not familiar with box scores, the columns for hitters are ab (at-bats), r (runs), h (hits), and bi (rbis or runs batted in). The columns for pitchers are ip (innings pitched), h (hits given up), r (runs scored), er (earned runs), bb (walks or base on balls), and so (strikeouts). Later, when you need to know, we'll say more about what an rbi is and when a run is an "earned run."

Jim: Ooh!

Can I tell who is up in the eighth?

Umm ... I don't think so. Total at-bats is 36 ... but that doesn't count walks ...

Hey, I'm getting nowhere.

Jerry: Look at the first inning instead.

Jim: Oh. Well, either Castro or Dominguez scores. Couldn't it be either one? I mean, let's see, first Abe is up. I guess he gets out. And then so does Brown. And then Castro could get on. And Dominguez could drive him in. So Castro could score in the first.

And could Dominguez score instead? Well sure, because Abe and Brown are out, then Castro ... can't get out, that would end the inning, ... so ... but if he gets on ...

Oh I see! Dominguez can't score in the first, because Abe, Brown, and Castro would all have to get out before Dominguez can score and that would end the inning!

So it must be Dominguez who scores in the eighth!

Jerry: And who drove him in?

Jim: Well, Engel of course. He's the only one who could.

Jerry: Couldn't Dominguez drive himself in?

Jim: Oh. Well. I guess he could. He could hit a home run.

Why are you looking at me like that?

Oh. No, he didn't hit a homer. If he did, it would say so at the bottom of the box score!
So Engel drove him in. And Dominguez drove in Castro in the first.
Wow.

That was tricky!

Jerry: Well, sort of.

But all you really needed was a basic understanding of the game.

Baseball has lots of rules. Most rules are so well-understood that you don't notice that you're using them. For example, in figuring out the puzzle, you understood that if Castro gets on base with 2 outs in the first inning, then Dominguez can't score. To score, without Castro getting out or scoring first, Dominguez would have to pass Castro on the basepaths. He would be called out for that.

Jim: Ah! Right. I'm sure I was thinking of that.

⟨pause⟩

Jim: But Jerry, what's the point? You can get the complete play-by-play summary of any game on mlb.com!

Jerry: But that misses the fun! The fun is in figuring out a puzzle!

For our puzzles, it's useful to know three sorts of rules:
(1) The rules of baseball.
(2) The rules used by scorers ruling on plays (hits, errors, …).
(3) The conventions governing how statistics appear in the box score (and these change over the years).

We assume the reader has a casual fan's knowledge of the game. When you need something more, it will appear in these pages.

By the way, the "Jim" and "Jerry" of this book are not us, they're not the authors. We just like the names.

—Jim and Jerry (the authors)

1.2 WORKING BACKWARDS

Jim: Jerry, I was thinking about that box score problem you showed me yesterday. For most innings the box score doesn't tell us which batters come to the plate, but we were able to make progress because we did know who leads off in the first inning. I was wondering if another way we could get started would be to work backwards from the end of the game. As you said, the box score gives us aggregate information. For example, if we know the number of at-bats for a team, won't that tell us who was the last batter on that team to hit in the game? Can we work this out from the at-bat statistics in the box score?

Jerry: You're on to something important there, Jim. It is possible to use aggregate statistics to figure out who was the last batter to hit in the game. But it's not through the at-bat statistics. They don't work for this purpose because they don't include every time a batter has his turn at the plate. For example, if a batter gets a walk, sacrifice fly, or sacrifice bunt, the plate appearance isn't counted as an official at-bat by the scorer, so it doesn't show up in that player's total number of at-bats.[2]

Jim: Oh, I see what you mean. I was thinking that the term "at-bat" referred to every time a hitter comes to the plate. But I guess that's not true.

Jerry: Call every time a player comes to the plate a "plate appearance." The total number of plate appearances is what we need to figure out who was the last guy to bat.

Jim: Okay, but I don't see any information about plate appearances in the box score, so what good does that do us?

---

[2]Generally, a sacrifice fly is a fly out in which a player on base scores after the ball is caught. It's only possible if there are fewer than 2 outs when the ball is hit. A sacrifice bunt is a bunt that enables a runner to advance and is not scored as a hit.

Jerry: Ah, but plate appearances are in the box score—they're just hidden. Every completed plate appearance either leads to (a) that batter making an out (either immediately or by being put out after getting on base), or
(b) scoring a run, or
(c) being left on base when the inning ends.
So for each inning, the total number of plate appearances equals the number of outs in the inning plus the number of runs scored in the inning plus the number of runners left on base in the inning. We can write this equation as:

$$PA = O + R + LOB.$$

That equation is valid for a single inning. It's also valid for the entire game, or indeed for any set of innings we want to examine.

If we're using the equation for the entire game, the box score gives us all of the values we need. The number of runs, R, is in the line score. The number of players a team leaves on base over the whole game, called "LOB," is given for each team just below the pitchers' statistics.

Jim: Ha! And the number of outs is 3 per inning, or 27 for the game, so ...

Jerry: Not quite! If the home team is leading after 8 1/2 innings, the game ends there and the home team has just 24 outs. Also, if the home team takes the lead in the ninth, the game instantly ends, with fewer than 3 outs in the inning (a "walk-off" win).

But the number of outs can always be found in the statistics for the pitchers of the opposing team. You look at the innings pitched, the ip column, for each pitcher. That's actually a measure of how many outs were made while the pitcher was pitching.

For example, if the pitcher pitched 2 2/3 innings, that means two full innings plus 2 outs, so the opposing team made 8 outs while he was pitching.[3]

Adding up the outs for each pitcher gives you the total number of outs in the game.

Here's a box score where you can make progress the way you suggested, working back from the end of the game:

Cats

	ab	r	h	bi
Abe ss	5	0	0	0
Brown 3b	4	0	1	0
Castro cf	4	1	1	0
Dominguez c	4	1	2	1
Engel rf	4	0	1	1
Frank 1b	4	0	0	0
Guzman lf	3	0	2	0
Hall 2b	4	0	0	0
Iglesias p	4	0	0	0
Totals	36	2	7	2

Dogs

	ab	r	h	bi
Amsler 1b	3	0	1	0
Blanco cf	3	0	0	0
Clark lf	4	0	1	0
Durand rf	3	0	1	0
Emerson 2b	2	0	0	0
Foster 3b	3	0	1	0
Gato c	3	1	2	1
Herrera ss	2	0	0	0
Ito p	3	0	1	0
Totals	26	1	7	1

```
Cats ··  0 0 1  0 0 0  0 0 1--2
Dogs ··  0 1 0  0 0 0  0 0 0--1
```

	ip	h	r	er	bb	so
Iglesias (W)	9	7	1	1	4	4
Ito (L)	9	7	2	2	1	3

LOB-Cats 8, Dogs 2, HR-Gato, 3b-Guzman, Durand
2b-Amsler, Foster, Guzman, Dominguez(2), SB-Emerson,
DP-Dogs 2

See if you can figure out who scored when for the Cats.

---

[3]In most box scores they don't use fractions for partial innings. Two and two-thirds innings pitched is usually written "2.2" and two and one-third innings is written "2.1". Sometimes it's written "2-2" and "2-1." It's odd, but printing fractions must have been difficult in the days of printing presses.

Jim: Okay, Jerry. I'll give it a try.

Well I see from the batting data that the Cats appear before the Dogs, so the Cats are the visiting team.[4] Since the game lasted 9 innings, and the Cats as visiting team completed the top half of the ninth inning, they must have had 27 outs—we don't even have to look at the Dogs' pitchers. But if we do, it gives the same result.

The box score has LOB: Cats 8, and the Cats scored 2 runs. So substituting into your equation, we have:

$$PA = 27 + 2 + 8 = 37.$$

Batting around the order four times makes for 36 plate appearances. For 37 appearances, the leadoff batter, Abe, must have been up one more time. Aha! Abe must have been the last batter to bat for the Cats.

Jim: Let's see, the number 3 and number 4 batters scored the two Cats runs. Whoever scored the run in the ninth had to bat in the ninth. But is that even possible when Abe is the last batter?

Jerry: Keep going.

Jim: Oh. Well, I guess I should try Dominguez, batting fourth. Let's see, if he bats in the ninth, then in that inning we would have batters 4,5,6,7,8,9, and 1 batting. That's seven men up in the 9th. In the 9th, there is 1 run and 3 outs, so looking at that equation, can PA be 7?

$$7 = 3 + 1 + LOB.$$

It's just possible. He would have to be driven in by Engel. And there would have to be three men left on base. That can happen.

---

[4]The visiting team is also listed first in the line scores (the inning-by-inning account of runs) and first in the pitching stats. The visiting team bats first in every inning.

And aha! Castro, in the lineup in slot 3 couldn't bat and score because that would require 8 plate appearances in the ninth.[5] Not possible, because you can't have more than three left on base in an inning! Therefore Castro must have scored his run in the third, driven in by Dominguez!

Jerry: Perfectly correct!

Now Jim, I have a different challenge for you. See if you can create a box score problem. Keep the same score with the Cats winning 2-1, but this time have them as the home team. You'll have to make some modifications to get it to work out right.

Jim: Ooh.

Worth remembering:

Plate appearances = Outs + Runs + Left On Base or more simply,

$$PA = O + R + LOB$$

1.3 A WALK-OFF

Jim: Of course I don't know what you were thinking when you challenged me to create a puzzle, but I've come up with something that I think works out. Here it is:

Dogs	ab	r	h	bi		Cats	ab	r	h	bi
Amsler 1b	3	0	1	0		Abe ss	4	0	0	0
Blanco cf	3	0	0	0		Brown 3b	4	0	1	0
Clark lf	4	0	1	0		Castro cf	4	1	1	0
Durand rf	3	0	1	0		Dominguez c	4	1	2	1
Emerson 2b	2	0	0	0		Engel rf	3	0	1	1
Foster 3b	3	0	1	0		Frank 1b	3	0	0	0
Gato c	3	1	2	1		Guzman lf	2	0	2	0
Herrera ss	2	0	0	0		Hall 2b	3	0	0	0
Ito p	3	0	1	0		Iglesias p	3	0	0	0
Totals	26	1	7	1		Totals	30	2	7	2

```
Dogs  ‥  0 1 0  0 0 0  0 0 0--1
Cats  ‥  0 0 1  0 0 0  0 0 1--2
```

	ip	h	r	er	bb	so
Ito (L)	8.2	7	2	2	1	3
Iglesias (W)	9	7	1	1	4	4

LOB-Dogs 2, Cats 3, HR-Gato, 3b-Guzman, Durand
2b-Amsler, Foster, Dominguez(2), SB-Emerson, DP-Dogs 2

---

[5]Think of the batting order as consisting of nine "slots." At every point in the game a team has nine players, each one batting in one of the slots.

Since it's the home team that wins, the game ends as soon as the Cats score their second run. It's a "walk-off" win, so there are fewer than 3 outs in the ninth inning for the Cats. There might be just 2 outs, 1 out, or even no outs. I chose to make it 2 outs so I gave the Dogs' pitcher '8.2' (8 2/3) innings pitched. And I also changed the LOB for the Cats to 3.

Jerry: Okay. Let me see if I can figure out who scored when. Our equation gives 26 outs + 2 runs + 3 LOB = 31 plate appearances. 27 plate appearances would be 3 times through the batting order, so Dominguez, batting in slot 4, must have had the last plate appearance.

As you said, the game ended only because the winning run had just scored, so Dominguez must have driven that run in. Dominguez didn't score the run in the ninth, because that could only happen if he hit a home run, and the box score shows no home run for him. So it must be Castro who scored the run.

Jim: You got it! Good problem, right?

Jerry: Wait a moment. Let's see what happened in the third inning. By elimination, we need to have Dominguez score and Engel drive him in. Can that happen?

Jim: Well why not?

Jerry: It can happen if they're up in the third inning. A puzzle isn't good if the conditions in the puzzle are impossible.

So let's just check to see if Dominguez and Engel could come to the plate in the third inning.

The Cats had a total of 3 LOB. That means that in the first 2 innings, by our equation,

$$PA = 6 \text{ (outs)} + 0 \text{ (runs)} + 0\text{–}3 \text{ (LOB)}$$

So there were anywhere from 6 to 9 plate appearances in the first 2 innings and leading off in the third was either Guzman, Hall, Iglesias, or Abe.

And we hope that Dominguez and Engel come to the plate in the third. I don't think that's possible.

Jim: But ... but ...

Jerry: For Dominguez to score the sole run of the inning he has to be one of the first 3 batters in the inning—because the guys ahead of him have to be out in order for him to score.

Jim: Oh!
Shoot.

Jerry: But here's an easy fix: Add 9 to the Cats' LOB, for a total of 12. That means every player has one more plate appearance (we can add 1 to each of the numbers in the Cats' players at-bat column). That doesn't change who comes to the plate in the ninth, so our analysis for the ninth inning is still valid. But it means that we can have anywhere from 0 to 6 LOB in the first 2 innings (the most you can have is 3 LOB in an inning because there are only three bases). Then it's easy for Dominguez and Engel to bat in the third inning. And now the puzzle works.

To make sure, we should construct a scorecard to see that the box score for the game is possible.

⟨minutes pass⟩

Here we go:

Cats	1	2	3	4	5	6	7	8	9	10	11	ab	r	h	bi
Abe ss	•	•		•		•		•				5	0	0	0
Brown 3B	1	•		•			•		•			5	0	1	0
Castro cf	•		•		•		•	①				5	1	1	0
Dominguez c	•		②		•		•	2				5	1	2	1
Engel rf	•		1		•		•					4	0	1	1
Frank 1B	•		•		•			•				4	0	0	0
Guzman lf		3	•			w		2				3	0	2	0
Hall 2B		•		•		•		•				4	0	0	0
Iglesias p		•		•		•		•				4	0	0	0
TOTALS												39	2	7	2

I used '1' for single, '2' for double, '3' for triple, 'w' for walk. A diamond is for a run. A dot is just a plate appearance, usually one where we don't know what kind—hit, out, etc. I used arrows to show when somebody bats somebody else in.

This is just one possibility. It shows how we could have a real game with the runs and rbis working out. As it is, it wouldn't produce 12 LOB, though, but that could easily be fixed by giving the Cats players some more hits in the box score. Nice puzzle, Jim!

Box Score Abbreviations:

ab - at-bats	r - runs	h - hits
bi - runs batted in	ip - innings pitched	er - earned runs
bb - base on balls	so - strike outs	LOB - left on base
HR - home runs	3b - triples	2b - doubles
SB - stolen bases	S - sacrifice (bunts)	SF - sacrifice flies
DP - double plays	WP - wild pitches	BK or B - balks
HBP - hit by pitch		

1.4 AN IMPERFECT GAME

Jerry: Jim, I have a new puzzle for you. Given the information in the following box score, can you tell me how the Cats scored the only run of the game?

Dogs

	ab	r	h	bi
Amsler 1b	4	0	1	0
Blanco cf	3	0	0	0
Clark lf	4	0	1	0
Durand rf	3	0	1	0
Emerson 2b	3	0	0	0
Foster 3b	2	0	1	0
Gato c	3	0	2	0
Herrera ss	2	0	0	0
Ito p	3	0	1	0
Totals	27	0	7	0

Cats

	ab	r	h	bi
Abe ss	3	0	0	0
Brown 3b	3	0	0	0
Castro cf	3	0	0	0
Dominguez c	3	0	0	0
Engel rf	3	0	0	0
Frank 1b	3	0	0	0
Guzman lf	3	0	0	0
Hall 2b	3	0	0	0
Iglesias p	3	1	1	0
Totals	27	1	1	0

```
Dogs ··  0 0 0   0 0 0   0 0 0--0
Cats ··  0 0 0   0 0 0   0 0 1--1
```

	ip	h	r	er	bb	so
Ito (L)	8.2	1	1	1	0	9
Iglesias (W)	9	7	0	0	4	4

LOB-Cats 0, Dogs 4, 3b-Iglesias, Durand
2b-Amsler, Foster, SB-Emerson, B-Ito,
DP--Cats 2

Jim: Okay, this looks kind of like the problem I made up, with a walk-off win in the ninth. I'll start by seeing who the last hitter was.

Our equation gives

$$PA = 26 + 1 + 0 = 27$$

—26 outs (from the 8.2 ip for the opposing pitcher), 1 run, and none left on base. That means the hitter in slot 9, Iglesias, was up last. It was a walk-off win, so he must have driven in the winning run, the only run of the game.

No, wait. Iglesias didn't have the rbi. He scored the run.

Hold it. Who drove … ?

Okay, this is crazy. Nobody drove in Iglesias.

So he didn't hit a homer. In fact I see he got a triple. Maybe that was in the ninth. I won't bet on it because you made this problem up.

And ... then ... he scored ...
And no one drove him in ...
So I guess he scored on an error ...

No. No errors in the game.

Hey, this is really nuts. Iglesias gets up in the ninth inning. He's the last guy up. Nobody is up after him. Somehow he scores![6]

Jerry: That's pretty good so far. Look a little more at the box score.

Jim: Ito pitched 8 2/3 innings, so he got two men out in the ninth before Iglesias scored. He gave up only one hit, that triple. He gave up no walks. Wow. This was practically a perfect game.

Wait. That "B–Ito." That's a balk, right? That's when the pitcher makes some sort of illegal move. When called by an umpire, the runner or runners move ahead one base. Aha! That's what happened! Iglesias triples. Then Ito balks, and Iglesias trots home, and the game is over!

Except they only call a balk on you when you pitch. In this case, that would have to be a pitch to Abe. But Abe didn't bat in the ninth! Iglesias had the last plate appearance.

As I said before, this is nuts!

Jerry: You've got it!

Jim: But what's going on? Abe didn't bat in the ninth. How do you balk if there's no batter?

---

[6]A run batted in (rbi) is credited to the batter who caused the run to score as a direct result of the batter's hit, walk, hit by pitch (HBP), or even an out. See the box at the end of the section for the one exception.

Jerry: Abe came to the plate, but it wasn't a "plate appearance" for the purposes of our equation, since it wasn't completed. And it wasn't an official at-bat. In the box score in the at-bat column (ab) only official at-bats are recorded. Coming up to the plate and even having some pitches thrown to you doesn't count as an at-bat, even though you literally did come to the plate.

For a balk, you need a man on base and a man at the plate, but you don't need a completed plate appearance. In this case, as soon as Iglesias crosses the plate, the game is over. The Cats win and Iglesias is the last person to have an official at-bat.

Another example. Say there are 2 outs and X is on first. Y comes to plate and takes a few pitches, then X is thrown out trying to steal second. That ends the inning. Y didn't finish his plate appearance. Y will lead off in the next inning with no balls or strikes and have a real plate appearance.

Jim: Sneaky problem!

Jerry: And you're not done! To solve this puzzle completely, you have to prove that your answer is correct.

Jim: Prove it?

Jerry: Yes. You've shown the run could have been scored this way, but is it the only way?

Jim: Hmm… A good puzzle should have one and only one answer. This is a good puzzle (you wouldn't give me anything second-class). So it has only one answer. And I found it. QED.

Jerry: No good! Thanks for the endorsement but this is a logic puzzle. It deserves a logical proof!

Jim: Arghh!

Okay. Well. What about arguing that there is no other way the run could be scored? I mean, no one made an error. The only mistake was the balk.

Jerry: Ah, but there is another way. Iglesias triples to center. The centerfielder tracks it down and throws the ball in to the cutoff man, say the second baseman. Iglesias rounds third base, thinking about getting an inside-the-park home run, but then stops short. The second baseman throws to the third baseman, to catch Iglesias off third base. The third base coach sees this and screams to Iglesias to run home. He does and he beats the throw home from the third baseman. Nobody makes a bad throw. Nobody drops the ball. No errors.[7]

Jim: Ooh.

Jerry: But you can prove this didn't happen.

Jim: Help! Help! Help! You said there's another way he could have scored! And now you say there isn't!

Jerry: Let's take that balk. There has to be a man on base for there to be a balk. When was that?

Jim: Ah! The only hit for the Cats is the triple. There are no walks. No hits. Nobody else can get on base!

Jerry: There are ways for a guy to get on base without a hit or a walk, but with one exception, these always lead to specific clues reported in the box score. You can get on if there's an error (E). You can get on if you're hit by a pitch (HBP).

A batter can get on base even if he strikes out if the catcher doesn't catch the third strike. When that happens he is allowed to run to first if either first base is empty or there are 2 outs. It's just the same as if you hit the ball into fair territory. You still have to reach first base without being tagged out and before a fielder holding the ball steps on first base.

[7]In baseball, "errors" are physical (bad throw, muffed catch, . . .). An error of judgement (throw to the wrong base, forget the number of outs, . . .) isn't recorded as an *error*.

Whenever somebody reaches first base on a third strike, it always leaves some trace in the box score. The official scorer is required to blame somebody. It might be on the pitcher for throwing a wild pitch (WP), the catcher, for not catching the pitch (a passed ball, or PB), or any fielder (usually the catcher or first baseman) for making an error (E) while trying to get the batter out before he reaches first base. At least one of these misplays gets reported in the box score.

But our box score doesn't report any of these clues (WP, PB, or E). There are no HBPs either, so nobody reached base in this game any of these ways.

That leaves just one other possible way to get on base. And it's a way that leaves no trace in the box score.

Jim: Yikes. Okay, what's that?

Jerry: You can reach base on a "fielder's choice." That's when you hit a ground ball and you could have been thrown out at first, but the fielder chooses to try to get another runner out instead, and that allows you to reach first base safely. That's not the case here, because you need a runner on base already for this. And there's no way here that anyone earlier could have reached base.

Jim: So we can conclude that nobody got on base at all until Iglesias triples. Then Ito balks. End of story. End of long story!

Scoring without an rbi:

If a player steals home, no one is awarded an rbi.
If an error, a passed ball, a wild pitch, or a balk enables a
     player to score on a play, no rbi is awarded.
If a batter grounds into a double play in which a run
     is scored, the batter does not get an rbi.

Getting on base:

	abbr.	noted in the box score	a trace in box score	counts as an at-bat	required:
hit	h	yes	yes	yes	
walk	bb	no	yes	no	
hit by a pitch	HBP	yes	yes	no	
fielder's choice	FC	no	no	yes	man on base
error	E	no	yes	yes	
...[8]					

1.5   THE FIVE-RUN INNING

Jerry: Another puzzle! It's fun making these up. Here's the box score:

Dogs

	ab	r	h	bi
Amsler 1b	5	1	1	3
Blanco cf	5	1	2	1
Clark lf	4	0	1	0
Durand rf	4	2	1	1
Emerson 2b	5	0	0	0
Foster 3b	4	0	1	0
Gato c	4	1	1	1
Herrera ss	3	2	1	0
Ito p	5	0	1	1
Totals	39	7	9	7

Cats

	ab	r	h	bi
Abe ss	4	1	1	2
Brown 3b	4	0	0	0
Castro cf	4	0	1	0
Dominguez c	3	0	0	0
Engel rf	4	0	0	0
Frank 1b	3	1	0	0
Guzman lf	3	0	0	0
Hall 2b	3	0	1	1
Iglesias p	2	1	1	0
Ivory p	1	0	0	0
Totals	31	3	4	3

```
Dogs ··  0 0 0  1 0 5  1 0 0--7
Cats ··  0 1 0  0 2 0  0 0 0--3
```

	ip	h	r	er	bb	so
Ito (W)	9	4	3	3	2	9
Iglesias (L)	6	8	6	6	6	4
Ivory	3	1	1	1	2	3

LOB- Dogs 13, Cats 3, HR-Amsler,
2b-Foster, SB-Emerson, DP-Cats 1

---

[8]There are a few more ways to get on base without an at-bat that the reader doesn't have to consider. These will be discussed in the appendix CRAZY THINGS THAT DON'T HAPPEN IN THIS BOOK.

And the challenge: For each of the Dogs' runs, find out when it was scored, who scored it, and who drove it in.

Jim: What a mess. 7 runs. 7 rbis. How can you analyze this? Can't *anyone* drive in anyone?

Jerry: Actually, no. In fact, the only people a batter can drive in, besides himself (with a home run), are the 5 batters that precede him—call this the "Five-step" rule. If one player, A is going to drive in another player, B, B first has to get on base. Then all the players between B and A must either get on base or be retired (out). There's room for only two more on the bases. And there's room for no more than two getting out (or else the inning is over). So there can't be more than four intervening players.

Jim: Okay. I'm going to make a chart of runs and drivers of runs.

players with rbis	number of rbis	guys who scored that he could drive in
Amsler	3	Amsler, Gato, Herrera (maybe twice)
Blanco	1	Amsler, Gato, Herrera
Durand	1	Blanco, Amsler, Herrera
Gato	1	Blanco, Durand
Ito	1	Gato, Herrera, Durand

I added Amsler to the players Amsler can drive in—Amsler is the only guy with a homer, so only he can drive in himself.

In this game, Durand scores twice. The only people who can drive him in are Gato and Ito, so Gato and Ito each drive in Durand. That takes care of Gato's rbis and Ito's rbis.

But that means that the only guy who can drive in Blanco is Durand, so Durand drives in Blanco. Hey! This is easy.

Amsler hit a home run. So Amsler drives in Amsler.
All that's left is Herrera's 2 runs and Gato's 1 run. Two of these are driven in by Amsler and one by Blanco.

player	rbis	definitely drives in	plus ...
Amsler	3	Amsler and Herrera	Gato or Herrera (again)
Blanco	1		Gato or Herrera
Durand	1	Blanco	
Gato	1	Durand	
Ito	1	Durand	

Does Amsler drive in Herrera twice or does he drive in Gato and Herrera each once?
Now I'm stuck.

Jerry: Look at the five-run inning.

Jim: Okay. 5 runs score in the sixth. That could be 1 run from each of the 5 who scored: Amsler, Blanco, Durand, Gato, and Herrera. Or maybe Durand or Herrera (the guys with 2 runs) scores twice in the inning? *Don't tell me the answer!!*

Okay, tell me the answer.

Jerry: If two guys score in the same inning, then all the intervening players must either score or be retired.

Jim: Okay, hold on. Here's the batting order along with who scores and who doesn't.

Amsler	scores
Blanco	scores
Clark	doesn't score
Durand	scores
Emerson	doesn't score
Foster	doesn't score
Gato	scores
Herrera	scores
Ito	doesn't score

Durand can't score twice in an inning because there are four guys who don't score. Same for all players. So all five scorers must actually score once in this inning to make five runs.

Then for the sixth inning, we need a string of batters which includes the 5 scoring players with no more than two non-scoring players in the middle of them.

The only way I see to do this is to somehow avoid Emerson and Foster. Let's see. Okay, then we should start with Gato and end with Durand. Then all five of the scorers can score and the guys in the middle who don't score are only Ito and Clark. They get out, but that's only 2 outs.

Whew!

Jim: Yeah. So the inning starts with Gato who gets on. Then Herrera gets on. He doesn't have an rbi, so Gato is still on. Then Ito is up. Does he drive in one of those guys? No. We know he drives in Durand. Not now, though. Some other time, I guess. Now he just gets out.

Next, Amsler is up and he has to score too. Does he hit his homer now? Yes, he must because if he hit his homer in another inning, he'd have 2 runs, but the box score says he has just one. So it's a 3-run homer, driving in Gato, Hernandez, and himself. So, going back to our earlier table, since Amsler drives in Gato, Blanco must drive in Hernandez. Not now, but sometime.

player	rbis	drives in
Amsler	3	Amsler, Gato, and Herrera
Blanco	1	Herrera
Durand	1	Blanco
Gato	1	Durand
Ito	1	Durand

That's 1 out and 3 runs in. Now Blanco gets on. Then Clark gets out. Then Durand drives in Blanco. We're done!

No we're not. Durand has to score too. He's going to be driven in by Gato—in Gato's second at-bat of the inning! But Emerson and Foster are up before Gato! We wanted to skip those guys!

But wait. There's only Durand on base, so there's room on the bases for the two non-scorers. So Emerson and Foster get on and the bases are loaded. And Durand is on third. Then Gato comes up for the second time in the inning and drives in Durand for the fifth run. Whew!

1st up in the sixth	Gato	gets on and scores
2nd	Herrera	gets on and scores
3rd	Ito	out
4th	Amsler	drives in G, H, and A
5th	Blanco	gets on and scores
6th	Clark	out
7th	Durand	gets on, drives in B, scores
8th	Emerson	gets on
9th	Foster	gets on
10th	Gato	gets on and drives in D

Jerry: Brilliant! Now what about the other 2 runs?

Jim: Ooh. The only guys left are Herrera (driven in by Blanco) and Durand (driven in by Ito). One of the two scores in the fourth inning, and one scores in the seventh.

Let's look at the seventh. Well, at the end of the sixth ... Umm...how does the sixth end? After Gato drove in Durand the bases are loaded. Well, one possibility is that Gato drove in Durand with a hit, and Emerson tried to score too but was thrown out at the plate. In that case Herrera would lead off the seventh.

If instead Herrera bats a second time in the sixth, he comes up with the bases loaded and 2 outs, so he would have to be the last batter of the inning (or else another run, a 6th, would have had to score). So either Herrera or Ito must lead off the seventh.

But wait! It can't be Ito leading off the seventh, because then none of the first 3 batters is Herrera or Durand, and there would be 3 outs before either guy could score. So Herrera must lead off the seventh, and he must be the one who scored in that inning, leaving Durand to score in the fourth. Driven in by Ito.

Done.

Jerry: That was complex, much more than I thought. But look, you didn't just come up with how the scoring might have gone. You actually *proved* that it must have gone this way, that the box score *forces* this solution.

To cap it off, we should make up a scorecard which shows that the box score is consistent.

DOGS	1	2	3	4	5	6	7	8	9	10	ab	R	H	rbi
Amsler 1B	•		•		w	HR	•		•		5	1	1	3
Blanco cf	•		•		1	1	1		•		5	1	2	1
Clark lf	1		•		w	•	•				4	0	1	0
Durand rf	DP			w	•	1		•			4	2	1	1
Emerson 2B		•		•	•	•		•			5	0	0	0
Foster 3B		w		1	•	•		•			4	0	1	0
Gato c		•		•		w	1		•		4	1	1	1
Herrera ss		•		•		1	w		w		3	2	1	0
Ito p		•		1	•	•		•			5	0	1	1
TOTALS	R 1 H			1 2		5 5	1 1				39	7	9	7

Jim: Done!

Jerry: *Well* done.

Jim: The "HR" is for home run and the "DP" is for double play. We didn't prove all the stuff here. The walks and hits are just there to show how it *could* have happened. What we proved is the runs and the rbis (arrows)

The Five-step Rule:

A player can only drive in himself and teammates who bat no more than five slots earlier.

The Gap Rule:
1. If batter **B** scores in an inning, each player batting before him in the inning either scores or makes an out before **B** scores.
2. If **A** and **B** both score in an inning, each player batting between **A** and **B** either scores or makes an out.
3. The first run of an inning must be scored by one of the first 3 batters of the inning. The second run of an inning must be scored by one of the first 4 batters of the inning. And so on.
4. If **B** scores twice in the same inning, his team must score at least 8 runs in that inning.[9]

1.6   CASEY AT THE BAT

Jim: Jerry, did you know that there's a magazine filled with baseball puzzles just like the ones we've been inventing?

---

[9]Any player **B** who scores is not allowed to pass any other runners. So everybody who bats before him must get out the way, either by being out or scoring before **B** scores. That's point 1. Point 2 is a special case of Point 1, since everybody batting between A and **B** bats before **B**. If **B** scores the first run of the inning, nobody batting before him can score, and at most 2 batters before him can be out, so **B** must be one for the first 3 batters. If **B** scores the second run of the inning, only one batter scores before him and at most 2 batters before him can be out, so **B** must be one of the first 4 batters. That's point 3. If **B** scores twice in an inning, there are 8 other batters who bat in between his two times up. All must score or be out. At most 2 can be out, so six must score. Adding in the 2 runs by **B** makes 8 runs. That's point 4.

Jerry: No, I didn't! That's amazing!

Jim: Well there isn't a magazine like that. But I did find a baseball problem in a math journal.

Jerry: Oh.

       Oh?

Jim: Here it is:

Casey, batting third for the Mudville Slugs, came to the plate in the ninth inning for his fourth at-bat. The bases were loaded. There were two outs, and his team was two runs behind. He struck out, ending the game. What was the final score?[10]

Solution on p. 181

---

[10]This puzzle first appeared in the authors' "Baseball Retrograde Analysis," *The Mathematical Intelligencer* 40(4): 71-76, 2018.

CHAPTER 2

Historical Mysteries

2.1 SUBSTITUTIONS

Jim: These puzzles are a lot of fun. I've enjoyed solving them and
        constructing one. But in the back of my mind is the thought
        that people really do play baseball. There are real box scores.
        They aren't made up. If we have a real box score can we ...?
        I mean, can we ... figure stuff out?

        You know, I'm almost afraid to ask, because that last
        example you showed me blew my mind, and I don't know if
        I'm ready for real life.[1]

Jerry: Well. It all depends on the game. Some box scores are quite
        simple, just like the first examples I showed you, and some
        are hopelessly complicated. But even with the messy
        games—the ones with so many runs and rbis that it's
        impossible to disentangle exactly how all the runs were
        scored—there are usually one or two interesting things one
        can deduce.

---

[1]To repeat, the "Jim" and "Jerry" in this book are not the authors. "Jerry," for
example, has a sister named Nancy in Oklahoma. Jerry, the co-author, also has a
sister Nancy, but she lives in Massachusetts.

DOI: 10.1201/9781003332602-2

One difference between real box scores and the examples I've shown you so far is that often real games, especially in recent years, have many player substitutions, and that makes for a long and messy box score. But, paradoxically, it can make it easier to figure out what happened in the game. Here's a pretty simple example, a game I picked more or less at random. It was played on July 19, 1988:

Cards 3, Dodgers 2

LOS ANGELES	ab	r	h	bi	ST. LOUIS	ab	r	h	bi
Sax 2b	5	0	2	0	Coleman lf	3	1	0	0
Heep 1b	2	1	0	1	OSmith ss	4	1	2	1
Gibson lf	4	0	2	0	Pndltn 3b	4	0	1	1
Marchal rf	4	0	2	1	Brnnsky rf	3	1	1	0
Shelby cf	4	0	0	0	Laga 1b	4	0	1	1
Scioscia c	3	0	1	0	Oquend cf	3	0	1	0
Dempsy c	1	0	0	0	TPena c	3	0	0	0
Hamltn 3b	4	0	1	0	Alicea 2b	2	0	1	0
Andesn ss	3	0	0	0	DeLeon p	2	0	0	0
Brennan p	2	0	0	0	Ford ph	0	0	0	0
Crews p	0	0	0	0	Quisnbry p	0	0	0	0
MiDavs ph	1	1	1	0	Dayley p	0	0	0	0
Holton p	0	0	0	0	Worrell p	0	0	0	0
Stubbs ph	1	0	0	0					
Totals	34	2	9	2	Totals	28	3	7	3

```
Los Angeles ........001  000  100—2
St. Louis ..........010  020  00x—3
```

Game Winning RBI—Smith (4).
DP—Los Angeles 2, St. Louis 1, LOB—Los Angeles 9, St. Louis 5. 2B—Scioscia, Davis. SB—Brunansky (13), Gibson (16), Coleman (47), Ford (4). SF—Heep.

Los Angeles	IP.	H.	R.	ER.	BB.	SO.
Brennan L, 0—1	4 2-3	6	3	3	3	2
Crews	1 1-3	1	0	0	0	0
Holton	2	0	0	0	1	1
St. Louis						
DeLeon W, 6—7	7	6	2	2	3	5
Quisnbry	0	1	0	0	0	0
Dayley	2-3	0	0	0	0	0
Worrell S,17	1 1-3	2	0	0	0	3

BK—Brennan. Umpires—Hallion, Williams, West, Engel. Time—2:47. Attendance—34,606.

Jim: Frightening.

Jerry: We can make sense of this.

Jim: The teams have more than nine players!

Jerry: There are nine slots. There are always just 9 batters in the batting order at any given time. When a simple substitution is made—one player replaces another—the new player takes the same slot in the batting order.

For each batting slot, the box score lists first the player who was originally in the game in that slot and then his replacements in the order in which they were put into the game. Sometimes a player who never bats at all is still listed. Look at the Dodgers' batters, and see if you can tell who was in which slot in the order.

Jim: OK, well everything looks simple for the first 5 batters. Except that Heep, playing first base, has only 2 ab (at-bats).

Jerry: He also has a sacrifice fly (SF). A sac fly is a plate appearance, but doesn't count as an at-bat. Walks also don't count as at-bats.

Jim: After these players, I see there are two catchers listed. Obviously, there can't be two catchers in the game at the same time. Scioscia must have been the starting catcher in slot 6, and he must have been replaced by Dempsy later in the game. Then we have Hamitn and Andesn batting in slots 7 and 8, followed by 5 more players who must all be batting in slot 9. Maybe that wasn't so hard.

- Sax 2b
- Heep 1b
- Gibson lf
- Marchal rf
- Shelby cf
- ⌈ Scioscia c
  ⌊ Dempsy c
- Hamitn 3b
- Andesn ss
- ⌈ Brennan p
  | Crews p
  | MiDavs ph
  | Holton p
  ⌊ Stubbs ph

Jerry: If you're worried about Heep, you can calculate how many plate appearances each slot in the lineup should have.

Jim: Oh. I can?

Jerry: How many plate appearances are there for LA?

Jim: Well, let's see, LA has 27 outs, 2 runs, and 9 LOB for a total of 38.

Ah. 38 is 9 times 4 plus 2. So every slot is up at least four times and the first two slots are up five times. Heep has 5 plate appearances but only 2 at-bats. One other slot is missing a plate appearance—slot 8, the shortstop.

But Cardinals pitcher DeLeon gave up 3 walks, so all the missing plate appearances are explained. Heep hit a sac fly and DeLeon walked him twice. And Deleon walked Andesn once.

Jerry: There are other differences between PA and ab. It doesn't count as an at-bat if you're hit by a pitch (HBP), make a sacrifice fly (SF) or a sacrifice bunt (S).

$$PA = ab + BB + HBP + SF + S$$

These all get mentioned in the box score. For this game we have

$$38 = 34 + 3 + 0 + 1 + 0.$$

The PA equation works for the whole team. It also works for a collection of innings or for an individual batting slot.

Jerry: Now see if you can figure out in which inning the pinch hitters in slot 9, Davs and Stubbs, batted.

Jim: Whoa! How can I do that?

Okay, hold it. The pitchers in this game all bat in slot 9, and we know from the pitching stats how long they were in the game. Brennan pitched for 4 2/3 innings, so he was replaced by Crews in the middle of the fifth inning. Crews pitched for 1 1/3, innings, so that takes us through 6 innings of pitching. He was removed from the game when Davs pinch-hit for him. But when was that, in the sixth inning or seventh inning?

Jerry: St. Louis was the home team (it comes second in the box score), so in each inning, the Dodgers bat first, and then the Cardinals. So after Crews pitched the sixth inning, the Dodgers batted in the seventh. So Davs pinch-hit in the seventh inning.

Jim: Okay. Then in the bottom of the seventh inning the Dodgers get a new pitcher, and it was Holton, who pitched the entire seventh and eighth innings—we know that because he had 2 full innings pitched, and besides, he was the last listed Dodgers' pitcher, so he had to finish out the game. But Stubbs is listed as a pinch hitter after Holton, so he must have pinch-hit in the top of the ninth inning. And then since the Cardinals were ahead after 8 1/2 innings, they didn't have to bat and the game was over, so LA didn't need any more pitching.

I have a question, though, Jerry. You said that sometimes a guy is listed as a batter even though he never appeared at the plate. It looks like this is the case for Crews and Holton, since they had 0 in the ab column. But how do we know they didn't come to the plate and get a walk, which wouldn't count as an at-bat?

Jerry: Good question! But you've already accounted for all the walks. No one was hit by a pitch, and Heep was the only player to make a sacrifice or a sacrifice fly. Another way to see this is that the other players in slot 9 already have 4 at-bats, which is just as many as the number of plate appearances you told me this slot should have.

Jim: Ah, Okay.

Hmm... the picture for St. Louis looks pretty easy. All the substitutions came in slot 9. They pinch-hit Ford for DeLeon, then put in Quisnbry to pitch the eighth inning. After he gave up a hit, he was replaced by Dayley who got 2 outs, and Worrell, who finished the game.

What's the "S,17"?

Jerry: The scorers awarded Worrell with a "save," meaning he came in when the team was ahead and held the lead to the end of the game. The "17" is there because it was his 17th save of the season.

Jim: What kind of a name is "Quisnbry"?

Jerry: In order to put in the box scores for all the major league games, papers leave out letters in some names. That's the kind of name "Quisnbry" probably is.

Jim: The Cardinals didn't play the ninth inning so they just had 24 outs. With 3 runs and 5 LOB that gives them 32 plate appearances. That means five players had 4 PA and the rest had 3.

Jerry: There are just 28 at-bats, but Dodger pitching gave up 4 walks, making up the difference. No one was hit by a pitch. So Coleman, Brnnsky, Oquend, and Pena got walks.

Now this was a real game. Neither of us made it up. But we can still figure out who scored, who drove in whom, and when. Try the Dodgers.

Jim: Okay. 2 runs. Heep and Davs scored. Heep and Marchal had rbis. Did Heep drive in Heep? Nope. Heep didn't homer. No one did. So Marchal drove in Heep. And so Heep drove in Davs.

Jim: When did each score? Scoring was in the third and seventh innings. Working backwards is sort of hard because it's hard to know how many players came up in the last 2 innings. And it's hard to work forwards because it's hard to know how many players came up in the first 2 innings!

Jerry: But you know something about Davs.

Jim: Oh. Yeah, he pinch-hit for Crews. Davs in the seventh inning, so Davs wasn't in the game in the third . . .

Jim: So we've got it! Marchal drove in Heep in the third and Heep drove in Davs in the seventh!

- Sax 2b
- Heep 1b
- Gibson lf
- Marchal rf
- Shelby cf
- Scioscia c
- Dempsy c
- Hamitn 3b
- Andesn ss
- Brennan p
- Crews p
- MiDavs ph
- Holton p
- Stubbs ph

Jerry: Great! Now what about the Cards?

Jim: Coleman, Smith, and Brnnsky scored, driven in by Smith, Pndltn (must be tough playing in the major leagues without a vowel), and Laga.

Coleman lf
OSmith ss
Pndltn 3b
Brnnsky rf
Laga 1b
Oquend cf
TPena c
Alicea 2b
DeLeon p
Ford ph
Quisnbry p
Dayley p
Worrell p

Any of Smith, Pndltn, and Laga could drive in Coleman. Either of Pndltn and Laga could drive in Smith. Ah! But only Laga can drive in Brnnsky, so he must drive in Brnnsky. And that leaves only Pndltn to drive in Smith, so he must drive in Smith. And that leaves Smith to drive in Coleman.

Jerry: But not necessarily in that order.

Jim: Oh yeah. Well, let's see who could come up in the second inning. No runs in the first, so anywhere from 3 to 6 batters came to the plate in the first. Brnnsky could come up in the second, if only 3 bat in the first. If 6 bat in the first, Coleman could come up.

But 3 guys (Pena, Alicea, and DeLeon) would have to be out before he scores the only run in the inning. Impossible. So Coleman and Smith can't score in the second.

So it has to be that Brnnsky leads off the second and scores when Laga … when Laga … hits? Hey, how is this going to work? Laga and Brnnsky have only singles! How can Brnnsky score?

Jerry: This is cool. Asking that question gets us to the heart of what we're doing! How Brnnsky scores is a difficult question. Even showing a single concrete way he could have scored requires an intimate understanding of the game. But you have already proved logically that he must have scored and that Laga must have driven him in! For the purposes of this deduction, it's irrelevant how he did it. For us as baseball detectives, the case is closed!

Coleman lf
OSmith ss
Pndltn 3b
Brnnsky rf
Laga 1b
Oquend cf
TPena c
Alicea 2b
DeLeon p
Ford ph
Quisnbry p
Dayley p
Worrell p

Jim: Okay.

Okay!

Okay. But I'm still curious . . .

Jerry: LA didn't commit any errors, so Brnnsky didn't score on an error.

But here's one way Brnnsky could have scored. Laga could have hit a towering fly ball so hard that he was sure it was a home run, and stopped to admire his hit. But the wind blew it back so it stayed in the park. It would have been a double if Laga had run hard, but delaying, Laga could only get to first, while Brnnsky, running hard all the way, scored.

And here's another way. As the box score notes, Brnnsky had a stolen base. He could have stolen second in the second inning and then have scored from second on Laga's single.[2]

> At-bats and Plate appearances:
>
> PA = ab + BB + HBP + S + SF

---

[2]For the record, Brunansky (real name) stole second in the second inning.

> Player Substitutions:
>
> For each batting slot, the box score lists the player who was originally in the game at that slot. His replacements at that slot are then listed in the order in which they came into the game. A player who never had a plate appearance (a pinch runner, for example) may still be listed.

2.2 CRAZY THINGS HAPPEN

> This section is about deductions that you can't make because of all the crazy things that can happen in a baseball game. Their existence helps explain why some things that seem obvious might not be true.
>
> It's fun to know what screwy things can happen. And surprisingly, you can sometimes prove that one of those crazy things must have happened.

Jim: Jerry, you've shown me some neat ways to make deductions, using the PA equations, the Gap Rule, and the Five-step Rule. You've shown me how we can sometimes tell how for every inning in a game with runs, who scored the runs, and who drove them in. But can we fill in more details? Can we say, what bases they occupied in between when they batted and scored? Are there any rules we can use to make this kind of deduction? And can we say anything about what happens in innings without any runs?

Jerry: Well, usually there's not much we can prove. For example, if a player gets a single, we can say that he safely reached first base for a moment. But that's basically all we can say. After that moment, just about anything could have happened, and we might have no way of knowing it.

Jim: What do you mean?

Jerry: Well, maybe he tried to reach second base and was thrown out. He'd still have credit for a single. Even if we could deduce he had a single in that inning, it doesn't necessarily follow that he stayed on first and ended up safe there.

In fact, he might have ended up at second, or third, or even scored on that same play, and we might have no way of knowing what happened.

Jim: That's crazy! If he got to second or third, wouldn't he be credited with a double or triple?

Jerry: Yes, it's crazy. But crazy things happen in baseball all the time. You only get credit for a double or triple if the reason you got to second or third base was how well you hit the ball. Sometimes you reach an extra base because of what the fielder did. He might have made an error that let you take another base. He might have thrown the ball to a different base trying to get another player out, and that let you get to second base.

Jim: Oh, I see. And he might have thrown to first base trying to get you out, and you took advantage of that to steam over to second base safely. And there wouldn't even have to be any errors scored on the play.

Jerry: That's right. I like to think about it this way: Just imagine all the crazy things that happen in a Little League game. They happen once in a while in the major leagues, too.

Jim: But does this mean we need to know a lot of baseball lore and strategy to know what is fair to deduce?

Jerry: Not really. All you have to do is avoid any deductions that aren't based strictly on the rules of baseball and scoring. Never assume that something is too crazy to happen if it's conceivable and not against the rules.

Here's a list of some crazy things that can happen:

1. A batter who hits a single can reach second or third, or even score on the play, even without an error. Or he may be put out on the play after reaching first base.

2. Similarly, a batter who reaches base on a walk or a third strike that is not caught can reach second, third, or home, even without an error, or may be put out on the play after reaching first base.

3. A batter can drive in a runner on first base with a single.

4. A batter can drive in multiple runners on a sacrifice fly.

5. A batter can get an rbi with a ground ball even if he doesn't get a hit and even if there are 2 outs, and even if no fielder is charged with an error.

6. A runner on base when another batter is at the plate can advance to another base, even without a stolen base being credited, without a balk, and without a wild pitch or passed ball.

Jerry: Here are two examples of this last crazy thing:

The catcher drops a pitch and the runner on first takes off for second base. The catcher throws to second base in time and the runner is caught in a rundown. He manages to evade a tag and slips out, reaching second safely. He wasn't trying to steal a base, so he doesn't get credit for a stolen base. The catcher threw to second in time, so it's not ruled a passed ball. There is no error.

A simpler example: with a player on first, the defense might concede second base by not holding the runner and not covering second base. If the man on first then runs to second on a pitch, there is no stolen base. It is judged "defensive indifference."

If any of these situations come up later, I'll explain them then!

2.3   THE GAME-WINNING RBI

Jim: Jerry, I think I've got the gist of how to analyze real box scores. I want to try one myself. Do you have one I could look at? Not too complicated. Just like the ones you gave me earlier, but maybe a little easier.

Jerry: Maybe! Here's another game from the same day:

A's 4, Indians 2

Cleveland	ab	r.	h.	bi	Oakland	ab	r.	h.	bi
Franco 2b	4	0	3	0	Javier lf	4	1	2	0
Francn 1b	4	0	0	0	DHedsn cf	4	2	2	2
Carter cf	4	0	0	0	Canseco rf	4	0	1	0
Kittle dh	4	1	1	0	McGwir 1b	4	1	1	2
Hall lf	4	1	2	1	Lansfrd 3b	3	0	0	0
Snyder rf	4	0	1	0	Baylor dh	3	0	0	0
Jacoby 3b	3	0	0	0	Hassey c	3	0	1	0
Bando c	3	0	1	0	Hubbrd 2b	3	0	0	0
Zuvella ss	3	0	0	0	Weiss ss	3	0	0	0
Totals	33	2	9	2	Totals	31	4	7	4

```
Cleveland .........0 1 0   0 0 0   0 0 1—2
Oakland ..........2 0 2   0 0 0   0 0 x—4
```

Game Winning RBI—McGwire (13).
  DP—Oakland 1, LOB—Cleveland 5, Oakland 3.
2B—Kittle. HR—McGwire (17), DHenderson(12),
Hall (3). SB—Canseco (23). SF—Jacoby .

Los Angeles	IP.	H.	R.	ER.	BB.	SO.
Swidell L, 10—9	8	7	4	4	0	8
Oakland						
Welch W, 11—6	8	7	1	1	0	8
Eckersley S,28	1	2	1	1	0	2

Umpires—Craft; Phillips; Morrison; Voltaggio.
Time—2:09.

Can you tell when each of Oakland's runs were scored and who drove them in? I don't think you'll have any trouble figuring it out.

Jim: I see that both Henderson and McGwire hit home runs. Henderson, batting second, scored twice. He drove himself in for 1 run when he homered. Since there were 4 runs and 4 rbis, every run must have been driven in by somebody. The only player besides Henderson who could drive in Henderson is McGwire.

That means that McGwire's 2 rbis are now accounted for—one for himself on his homer and one for driving in Henderson. And that leaves only one more rbi, the second rbi for Henderson. He must have driven in Javier.

So that's who drove in whom. Now, when? Henderson can't score his 2 runs in one inning. He must have scored in both the first and third innings, the only times that Oakland scored. The other runs have to be scored by McGuire and Javier, one of them in the first inning, and the other one in the third.

Hmm... and does Henderson homer in the first or the third? And when he drives in Javier, is it with the homer or a different hit?

What do I do now?

Wait wait wait. It says in the box score, "Game Winning RBI—McGwire!" So I guess that McGwire drove in ... drove in ... drove in ... Hey, which run is the game-winner? Cleveland got 2 runs, so for Oakland to win, it had to score at least 3 runs. So is the third run the game-winner?

Jerry: Um, no. Officially, the "game-winner" is the run that puts the team ahead for good. In this case, that's the first run of the game. After the first inning, Oakland has the lead and never loses it.[3]

---

[3]Game Winning RBIs were added to box scores in 1980. And they were removed in 1989. The shape and content of box scores changes through the years, sometimes radically.

Jim: Ooh. Okay, then Henderson can't drive in Javier in the first because if he did, that would be the game-winning rbi. And he can't hit his homer in the first for the same reason. But he does score. Two people score, necessarily Henderson and McGwire. The only way that can happen is if McGwire hits his homer in the first inning, driving in Henderson and himself. And that means that Javier and Henderson score in the third, on Henderson's homer. Got it!

Jerry: Perfect!

Jim: Next, the Cleveland side. Is that going to be easy, too?

Jerry: Actually, I haven't looked at that yet. Let's see what we can do.

Cleveland

	ab	r	h	bi
Franco 2b	4	0	3	0
Francn 1b	4	0	0	0
Carter cf	4	0	0	0
Kittle dh	4	1	1	0
Hall lf	4	1	2	1
Snyder rf	4	0	1	0
Jacoby 3b	3	0	0	0
Bando c	3	0	1	0
Zuvella ss	3	0	0	0

Jim: Well to start, I see that Hall had a home run. That accounts for 1 run and 1 rbi. And I see that Jacoby had a sacrifice fly (SF), so he must have driven in the other run, scored by Kittle. Hey, but why does Jacoby's line show no rbi? How can that be? If you have a sacrifice fly, you drove in a run, right?

Jerry: Yes, you're right. It's a typo in the box score![4]

Jim: Hmm. Looking just at the second inning it appears either Kittle or Hall could have scored. But what about the ninth? Using the PA equation, we have PA = 27 outs + 2 runs + 5 LOB = 34 = 27 + 7. That means that Jacoby, batting seventh, was the last batter for Cleveland in the ninth inning.

---

[4]Newspapers in those days printed many box scores every day. A typo isn't surprising, but in all the box scores we used for this book, this is the only one where we saw a mistake.

Jerry: That makes it unlikely that that's the inning in which Jacoby hits a sacrifice fly.

Jim: It makes it impossible! You can't do that in the last at-bat of an inning, right? When Jacoby's fly ball was caught, there couldn't be 3 outs, or else the inning would have been over and Kittle couldn't have scored. So in that case there would have to be another batter up. And that's impossible. That means the run in the ninth must have been Hall's homer!

Jerry: Wait a moment. Let's consider your argument. Why do you think there would have to be another batter after Jacoby?

Jim: Well, after the run scored, the inning isn't over yet, because there are only 2 outs. So, of course there must be … Oh, I see. Maybe there was another runner on base, say on second base, and he tried to advance on the sacrifice fly, and was thrown out at third base!

Jerry: Yes, that would work. Another possibility is that another batter came to the plate after Jacoby, but before he could complete his plate appearance, a runner was picked off base for the third out.

Jim: So there isn't any way to know in which inning which run was scored!

Jerry: I'm not ready to give up! Let's look for more clues in the box score.

Jim: More clues? What clues?

Could we show, maybe, that there couldn't be another runner who could have been on base to be thrown out? Let's see. After Kittle got on base, Hall or Snyder could have had a hit and been on base. The box score shows that Hall had 2 hits and Snyder, batting after him, also had a hit …

Umm… is there anything in the pitching stats?

Los Angeles	IP.	H.	R.	ER.	BB.	SO.
Swidell L, 10—9	8	7	4	4	0	8
Oakland						
Welch W, 11—6	8	7	1	1	0	8
Eckersley S,28	1	2	1	1	0	2

Eckersley pitched the ninth and gave up 2 hits, and the run he gave up was earned. But the 2 hits could have been a homer by Hall and a single by Snyder or Jacoby.

Or it could be hits by Kittle and Hall.

That doesn't help at all.

Jerry: I got it.

Eckersley only pitched the ninth inning. He had 2 strikeouts, and since he had no wild pitch, and there were no errors or passed balls in the game, the batters who were struck out couldn't have reached base on an uncaught third strike. Now Jacoby had the last plate appearance in the game. If Jacoby had a sacrifice fly in the ninth, he didn't strike out. So the 2 strikeouts came before his at-bat. But there can be at most 1 out before a sacrifice fly. So Jacoby's sacrifice fly couldn't have been in the ninth. It had to be in the second. Hall homered in the ninth. Problem solved.

Jim: Wow.

The winning …

The **game-winning rbi** was, in the 1980s, awarded to the player that drove in the run that put his team ahead for good. Similarly, the **winning pitcher** is the player who was the pitcher of record when his team took a lead that it never relinquished.

2.4 NUMBER 299

Jerry: Jim, I'm working on an interesting box score problem, and I wonder if you'd be interested in helping me with it.

Jim: Help you? It's hard enough just following what you're thinking! But sure, I'm happy to take a look.

Jerry: Here it is, a game played on Sept. 8, 1962. I'm trying to figure out when each White Sox batter scored their runs, and who drove them in.

CHISOX WIN

WASHINGTON	ab	r	h	bi	CHICAGO	ab	r	h	bi
Piersall,cf	5	1	1	0	Aparicio,ss	4	1	2	1
Cottier,2b	4	1	1	0	Cun'gham,1b	5	1	1	0
Hinton,rf	4	1	2	1	Robinson,lf	5	1	2	2
Zipfel,1b	4	0	1	0	Landis,cf	3	2	2	0
Retzer,c	3	0	2	1	Hershb'ger,rf	4	0	0	0
Johnson,3b	4	0	0	0	Fox,2b	2	0	0	0
Lock,lf	4	0	1	1	Kenworthy,2b	1	0	0	0
Kennedy,ss	4	0	0	0	cJones	1	0	1	1
Rudolph,p	1	0	0	0	Martin,3b	0	0	0	0
aHicks	1	0	0	0	Carreon,c	3	0	1	2
Burnside,p	0	0	0	0	Esp'sito,3b,2b	3	1	1	0
bO'Connell	0	0	0	0	Wynn,p	3	0	0	0
Hannan,p	0	0	0	0					
dKing	1	0	1	0	Total		34 6 10 6		
Total		35 3 9 3							

aGrounded out for Rudolph in the 5th; bWalked for
Burnside in the 7th; cSingled for Kenworthy in 7th;
dSingled for Hannan in 9th.

Washington ....2 0 0   0 0 1   0 0 0—3

Chicago ....0 0 0   4 0 1   1 0 x—6

E—Johnson 2. A—Washington 10, Chicago 11. DP—
Kennedy, Cottier, Zipfel; Kenworthy, Aparicio, Cunning-
ham, LOB—Washington 7, Chicago 9. 2B—Carreon,
Aparicio. SB—Hinton, Landis

	IP.	H.	R.	ER.	BB.	SO.
Rudolf (L, 8—8)	4	4	4	4	3	1
Burnside	2	4	1	1	1	0
Hannan	2	2	1	1	1	1
Wynn (W, 7—12)	9	9	3	3	2	3

Umpires—Napp, Schwarts, Stevens, Drummond.
Time—2.29. Attendance—3,679.

Jim: OK, following your methods, We should first sort out Chicago's lineup, right? Let's see, Chicago has 6 runs + 24 outs + 9 LOB = 39 PA—that makes 5 plate appearances for the first 3 slots in the order and 4 plate appearances for the others ... Chicago has 34 at-bats ... and there are 5 walks given up by the 3 Washington pitchers. Perfect—that accounts for all 39 PA. Aparicio must have had a walk to bring him up to 5 PA, and Landis had one to bring him up to 4.

Next, Fox would need 2 walks, and ... no, that can't be right, because there's another second baseman right after him who must be in the same slot in the batting order, and then a pinch hitter, Jones. All of them must have been batting in slot 6 in the order, and together their 4 at-bats account for all 4 plate appearances in that slot.

Hey look—that letter c before Jones isn't a first initial, it's a footnote mark, and below it says that he singled in the seventh inning. So, so, so, since the other players in slot 6 have 3 of the 4 PA, Jones only batted once, so his rbi must have been when he singled in the seventh.

Jerry: You won't believe this, Jim, but you've helped me already! I was so eager to jump ahead to figure out the fourth inning that I skipped the basics. I just wrote down which slots could drive in other slots and ignored the clue about when Jones got his rbi!

Let me tell you what I worked out for the fourth inning—

Jim: Whoa, Jerry. Shouldn't we finish the basics? Let's see who can drive in whom. I want to list the slots, who scores, and the rbis, and work this out. Like you showed me before.

player	runs	rbis	could drive in
Aparicio	1	1	Esposito
Cunningham[5]	1		
Robinson	1	2	Cunningham, Aparicio
Landis	2		
Hershberger			
Jones		1	Landis, Robinson, Cunningham
Carreon		2	Landis, Robinson
Esposito	1		
Wynn			

[5]I'm restoring retired ball players' missing letters -J.

No home runs, so nobody drove in themselves. Looking five steps back from Aparicio's slot in the lineup, only Esposito has a run, so Aparicio drove in Esposito. Now that Esposito's run is accounted for, Robinson only has Aparicio and Cunningham in range so he must have driven them in. That leaves only Robinson's run and Landis' 2 runs. They must be driven in by some combination of Jones and Carreon.

player	drove in
Aparicio	Esposito
Robinson	Aparicio and Cunningham
Jones	Robinson or Landis
Carreon	Robinson and Landis *or* Landis twice

Okay, *now* tell me what you figured out for the fourth inning.

Jerry: Okay. But let me start by saying that in figuring out this box score, I used just about every trick I've taught you so far! It's that tricky.

So in the fourth inning, 4 runs scored. They could be scored by any four of the five who score in the game. But by the Gap Rule (p. 24), no more than two non-scorers can bat before anybody who scores in the inning, so we can't have all three non-scoring slots, 5, 6, and 7, bat before someone who scores. So the 4 runs have to follow this order:

Esposito–Aparicio–Cunningham–Robinson–Landis.

Jim: Okay. I get that. So who's the guy who doesn't score in the fourth?

Jerry: With your discovery about Jones, I think it's Esposito. But tell me what you think of this reasoning.

Jim: Okay.

Jerry: If Esposito scores he has to be the first (to avoid slots 5, 6, and 7). He's driven in by Aparicio. Now, who can drive in 3 more runs? Robinson can drive in two. But the third has to be driven in by Carreon, since Jones doesn't enter the game until the seventh inning. That means at least 9 players up in the fourth inning.

Jim: That can happen, right? With 4 runs, 3 left on base, 3 outs, you can have as many as $4 + 3 + 3 = 10$ plate appearances in an inning.

Jerry: Yeah, but ... well, let me show you where that leads.

In the first 4 innings, there are a total of 4 runs, 12 outs, and anywhere from 0 to 9 left on base (Chicago has 9 LOB for the entire game) for a total of 16 to 25 PA. Carreon bats in slot 7, so he is either the 16th or the 25th batter of the game when he bats in the fourth inning. He can't be the 16th batter, because with at least 9 batters up in the fourth, that would leave only $16 - 9 = 7$ batters for the first 3 innings.

So Carreon must be the 25th batter in the game when he comes up in the fourth. That means that all 9 LOB must be used up in the first 4 innings. That leaves no LOB for the rest of the game!

Jim: That doesn't sound good. But can you prove that it's impossible?

Jerry: With no LOB in the late innings, we work backwards. Chicago only batted in 8 innings. So (with all the LOB used up) there were just 3 batters in the eighth inning (no runs) and since the last batter was in slot 3, that means Aparicio, Cunningham, and Robinson in the eighth. Then there were exactly 4 batters in the seventh inning (1 run), namely, Jones, Carreon, Esposito, and Wynn. But if Esposito scored in the fourth, then none of those 4 batters can score in the seventh! That contradicts the box score!

Jim: Okay. You're right then. Esposito couldn't have scored in the fourth inning. Then it must have been Aparicio, Cunningham, Robinson, and Landis, in that order.

Jerry: Now who scored in the seventh? The only runs left to place are Esposito's and Landis'.

Jim: Well we know that Jones drove in the run in the seventh. But Jones can only drive in Robinson or Landis and Robinson's only run was in the fourth, so Jones must have driven in Landis. And that means Esposito scored in the sixth, driven in by Aparicio. And that also means that in the fourth inning it was Robinson who drove in both Aparicio and Cunningham, while Carreon drove in Robinson and Landis.

player	drove in
Aparicio	Esposito
Robinson	Aparicio and Cunningham
Jones	Landis
Carreon	Robinson and Landis

Jerry: That's it! That's how all the runs took place!

I think it's helpful to draw a scorecard to make sure this all fits together.

⟨minutes pass⟩

Chicago	1	2	3	4	5	6	7	8	9	10	11	ab	R	H	rbi
Aparicio	•				•		•					4	1	2	1
Cunningham	•				•		•					5	1	1	0
Robinson	•				•		•	•				5	1	2	2
Landis		•			•		•					3	2	2	0
Hershberger		•		•			•					4	0	0	0
Fo/Ke/Jo/Ma	•			•		•	1					4	0	1	1
Carreon			•	•		•	•					3	0	1	2
Esposito			•	•			•					3	1	1	0
Wynn		•		•	•		•					3	0	0	0
TOTALS R/H				4		1	1					34	6	10	6

Okay, it all works!

And in fact, since Aparicio has to lead off the fourth, there can't be any LOB in the first 3 innings:

So after a very smooth start, the Washington pitcher Rudolph suddenly lost his stuff, lost his lead, and was removed from the game.

Jerry: And on the other side, Wynn righted himself after a rocky first inning, and earned the win.

I haven't told you this, but this was the 299th win of Wynn's career—that's the immortal Early Wynn. And it came just two days before Mickey Mantle's 400th home run.

2.5   NUMBER 400

Jim: Jerry, I looked up that game, the game where Mantle hit his 400th homer. The box score doesn't say whether he homered in the fifth or ninth inning. Figuring out the timing and details—which inning, who was on base, how many outs there were—made for a good puzzle.

Jerry: I do like good puzzles.

Solution on p. 181

NEW YORK					DETROIT				
	ab	r	h	bi		ab	r	h	bi
Kubek,ss	4	0	0	0	Fernandez	4	0	1	0
Richardson,2b	4	1	2	0	Bruton,cf	4	0	1	0
Tresh,lf	3	0	0	0	Kaline,rf	3	1	2	1
Mantle,cf	3	2	2	1	Colavito,lf	4	0	1	0
Lopez,rf	4	0	1	1	cWood	0	0	0	0
aLinz	0	0	0	0	Cash,1b	3	0	0	0
Maris,rf	0	0	0	0	bMorton	1	0	0	0
Howard,c	4	0	2	1	McAuliffe,2b	3	0	1	0
Skowron,1b	4	0	0	0	Kostro,3b	3	0	0	0
Boyer,3b	3	0	0	0	Brown,c	3	0	0	0
Terry,p	3	0	0	0	Aguirre,p	3	0	0	0
Daley,p	0	0	0	0	Fox,p	0	0	0	0
Total	32	3	7	3	Total	31	1	6	1

aRan for Lopez in the 9th; bHit into force play for Cash in 9th; cRan for Colavito in 9th.

New York ....0 0 0   0 1 0   0 0 2—3
Detroit ....1 0 0   0 0 0   0 0 0—1

E—None 2. A—New York 14, Detroit 11. DP—Boyer, Richardson, and Skowron; McAuliffe, Fernandez, and Cash. LOB—New York 4, Detroit 4. 2B—Richardson, Howard. HR—Mantle, Kaline. SB—Fernandez. S—Tresh.

	IP.	H.	R.	ER.	BB.	SO.
Terry (W, 21—10)	8.2	6	1	1	1	5
Daley	0.1	0	0	0	0	0
Aguirre (L, 14—7)	8.1	7	3	3	1	8
Fox	0.2	0	0	0	0	0

Umpires—Smith, Rice, Paparella, Scar. Time—2.05. Attendance—22,810.

2.6 WHO SCORED WHEN

Jerry: In the box score below[6] I was able to figure out which Mets scored in which innings. Never mind the Cardinals. Try it.[7]

```
                    FIRST GAME
        ST. LOUIS              METS
            ab r. h. bi            ab r. h. bi
Coleman, lf  5 0 0 0  Miller, 2b    4 0 1 2
O. Smith, ss 4 0 1 1  Samuel, cf    3 0 1 0
M. Thmp, cf  3 0 0 0  H. Jhnsn, 3b  3 1 1 0
Guerrer, 1b  4 0 2 0  Strwbry, rf   4 0 0 0
Pndltn, 3b   4 0 1 0  McRylds, lf   3 2 1 1
Oquend, 2b   3 0 1 0  Magadn, 1b    3 1 1 0
Brnnsky, rf  4 0 1 0  Teufel, 1b    0 0 0 1
Dayley, p    0 0 0 0  Sasser, c     3 0 0 0
Quisnbry p   0 0 0 0  Carter, c     1 0 1 1
T. Pena, c   2 1 0 0  Elster, ss    2 1 0 0
Walling, ph  1 0 0 0  Darling, p    3 0 1 0
Pagnozzic, c 1 0 0 0  Totals       29 5 7 5
DeLeon, p    2 0 1 0
Morris, rf   1 0 0 0
Totals      34 1 7 1
St. Louis ..............001  000  000—1
Mets ...................030  000  02x—5
```

E—Miller. LOB—St. Louis 9, Mets 6. 2b—Miller, O.Smith, H.Johnson. HR—McReynolds (13). SB—Oquendo (2), McReynolds (13). S—DeLeon. SF—Teufel.

St. Louis	IP.	H.	R.	ER.	BB.	SO.
DeLeon L. 11-11...	7.1	6	5	5	4	8
Dayley.............	0.1	0	0	0	0	0
Quisnbry...........	0.1	1	0	0	0	0
Mets						
Darling W. 10-9....	9	7	1	0	1	3

HBP—Samuel by DeLeon, M. Thompson by Darling. BK—Darling. Umpires—Home, Marsh; First, Hohn; Second, Wendelstedt; Third, Darling. T—2:49.

Jim: I will!

Solution on p. 182

---

[6] August 10, 1989

[7] This puzzle first appeared in the authors' "Retrograde Baseball Redux," *The Mathematical Intelligencer* 44(4), 2022.

2.7   EARNING AND UNEARNING

Jerry: My sister Nancy called me yesterday and asked me what I've been up to. I told her about these puzzles. She asked for one and I emailed her this box score I found from June 14, 1970.

CHICAGO	ab	r.	h.	bi	LOS ANGELES	ab	r.	h.	bi
Kessinger, ss	4	0	0	0	Wills, ss	3	1	0	0
Popovich, 2b	4	1	3	1	Mota, lf	3	0	1	1
Williams, lf	4	0	0	0	Davis, cf	3	0	0	0
Hickman, rf	5	1	2	2	Parker, 1b	4	0	0	0
James, cf	0	0	0	0	Kosco, rf	4	1	1	0
Calison, rf	4	0	1	0	Gr'b'k'witz, 3b	4	0	1	0
Sento, 3b	3	0	0	0	Haller, c	4	1	1	1
Banks, 1b	4	1	1	1	Joshua, pr	0	1	0	0
Hiatt, c	3	1	0	0	Lefebvre, 2b	3	0	1	0
Decker, p	1	0	0	0	Singer, p	0	0	0	0
Regan, p	1	0	0	0	Gabrielson, ph	1	0	0	1
					Pena, p	0	0	0	0
					Crawford, ph	1	0	1	0
					Brewer, p	0	0	0	0
Total	33	4	7	4	Sudakis, ph	1	1	1	2
					Total	31	5	7	5

Chicago .... 2 2 0   0 0 0   0 0 0—4
Los Angeles .... 1 2 0   0 0 0   0 0 2—5

E—Decker. LOB—Chicago 10, Los Angeles 5. 2B—Kosco. HR—Hickman (16), Banks (6), Sudakis (4). SB—James. S—Decker, Davis. SF—Mota.

	IP.	H.	R.	ER.	BB.	SO.
Decker	6.2	4	3	2	3	3
Regan L (4-2)	*2	3	2	2	0	0
Singer	2	3	4	4	2	1
Pena	5	3	0	0	3	1
Brewer W (2-1)	2	1	0	0	1	3

*Two out when winning run was scored.
HBP—by Singer (Popovich)
Time—2.34. Attendance—28,756.

Jim: Is she a fan?

Jerry: No. But she knows stuff. I'm always surprised at what she knows. I told her about the equation for plate appearances and she quickly figured out who scored when and who drove in whom.

Jerry: Nancy didn't know what "earned runs" were, but she noticed that there were columns for "ER" and "R" and asked about it. She noticed that in Decker's line the numbers in these columns were different. So I explained about earned runs and unearned runs. And I told her the unearned run must have been the one scored by Wills. Wills didn't have a hit, but he could have walked. And then he had to be driven in by Mota. But Mota has a sacrifice fly, so that's how Mota drove him in. I was really getting into it. She's my younger sister, you know, and she looks up to me. I have a reputation to keep up.

Jim: Good for you!

Jerry: So I said: "So how can you score from first on a sacrifice fly? That's crazy. And there's no stolen base in the box score, so he couldn't have got to second base. But there is an error by the pitcher, Decker. So somehow Wills must have gotten to third base on an error by the pitcher, where he could score on the sacrifice fly. And that's why the run was unearned.

Jim: But Jerry, don't you keep reminding me that crazy things can happen? Things like advancing bases without any trace in the box score? You can't really know for sure there had to be an error then, can you?

Jerry: I know, I know! And wouldn't you know, Nancy challenged me on that! "How is that a proof?" she wanted to know. "I thought you wanted strictly logical reasoning in these problems." She nailed me. She wasn't mean about it, but I was so embarrassed, I didn't want to give in. I looked up the play-by-play on the Web, and there it was—Wills really did get to third on the pitcher's error.

Jim: But you knew she was right.

Jerry: Yeah.

Jim: . . . uh, Jerry, now that you mention it, exactly how would you define an earned run?

Jerry: An unearned run is a run that was only scored because of an error or a passed ball. The pitcher didn't really "earn" the run, that is, he didn't deserve the black mark on his pitching record. It was a fielder's fault that it was scored, not his fault as the pitcher.

Jim: Can you give me an example?

Jerry: Sure. A common way—the simplest example—is when a batter gets on base due to an error and later scores a run. He should have been out, so it's not the pitcher's fault he scored. It's still charged as a "run" to the pitcher,[8] because he was pitching when it happened but it wasn't an "earned run." It's an "unearned run."

Jim: So Wills got on base due to an error, and that's why his run was unearned?

Jerry: Not quite. In the actual game, he walked. Now Wills was a gifted base stealer and a daring runner. He took a big lead off first base and Decker threw to first base for a pickoff. But Decker threw the ball wildly, and Wills ran all the way to third. A good pickoff throw would have put Wills out, so the run was unearned.

Jim: Wait a minute! It was still the pitcher's fault, wasn't it? He threw the ball away. He was the one charged with the error. So it should be an earned run, shouldn't it?

Jerry: It's hard to disagree with you on that. Maybe it ought to be called an earned run. But it's not! Earned and unearned runs are a measure of the pitcher's pitching performance, not his fielding performance. Decker-the-pitcher wouldn't have given up a run if his alter ego, Decker-the-fielder, had made an accurate throw to first.

---

[8]We are assuming here that there is only one pitcher in the inning. If there are multiple pitchers in an inning, the rules for who is responsible for which runs are more complicated.

Jim: Oh. All right then. I have another question. You said that the official scorer ruled that a good throw would have put Wills out at first. But what if he thought that Wills would have been safe at first even with a good throw? Then would the run be earned?

Jerry: That's a tougher question! After Wills was up, in the actual game, Mota had a sacrifice fly, Davis had a groundout, and Parker struck out. Now, in your scenario, Wills should have still been on first base when Mota came up to bat.

The official scorer's job is to reconstruct how the inning would have proceeded in that case. There might be some judgment calls involved. But he probably would have said that Wills would have stayed on first on Mota's fly ball. Maybe he'd think Wills could have reached second base on the ground ball. But the strikeout would have ended the inning with no runs. So he still would have ruled it an unearned run.

> How were the other runs scored? Can you come up with a scenario whereby one of the other runs was unearned?

Solution on p. 184

> **Unearned runs:**
>
> If a run scores due to an error or passed ball, it's called an "unearned run." More precisely, the official scorer reconstructs how the inning would have proceeded without any errors or passed balls. All runs that would have scored are "earned." All others are "unearned." Here are two examples of unearned runs:
> (1) Any run scored by a player who reached base due to an error (even the pitcher's error), or who would have been put out on the bases but for an error.
> (2) Any run scored after there should have been 3 outs, assuming no errors or passed balls were made.

CHAPTER 3

Steady Games

3.1   A STEADY GAME

Jim: Jerry, I have an idea for a puzzle. It's a little odd. But suppose that in one game, the same number of players came to the plate every inning. And the game had this line score:

Team A ······ ??? ??? ???
Team B ...... ??? ??? ??x

The question is: Team B won, but did Team A score any runs?

Jerry: You mean, for example, if the total number of plate appearances in the first inning, some with Team A and some with Team B, was 8, then in every inning there would be 8 batters? But maybe in one inning A would have 5 batters and B would have 3, and in another each team would have 4 batters?

Jim: Yup. You got it.

Jerry: Hmm ...

Well, of course! It's pretty easy.

Solution on p. 186

DOI: 10.1201/9781003332602-3

3.2 ANOTHER STEADY GAME

Jim: I've got another one. This time the home team has the same number of plate appearances each inning, but the number of plate appearances for the visiting team might vary. The game was 9 innings. Team A, the home team, scored 2 runs and left 7 players on base. Who won the game?

Jerry: Okay. That's *your* question. My question is: What's the story? What's your excuse for this problem and the last problem?

Jim: Yeah, I've been thinking about that. Maybe one team is so much better than the other that they can engineer games like this. Would that work?

Jerry: Maybe. Maybe it's a team of U.S. players in a country with no tradition of baseball.

Jim: How about Italy?

Jerry: Yeah, well, okay, but I'll bet there are some good players there. After all, they must know all about Italian-American stars like Justin Verlander, Aaron Nola, and Nathan Eovaldi ...

Jim: Then let's make it a women's team. The Brooklyn Bombshells. They're barnstorming across Italy playing women's teams formed for the occasion. The Bombshells are Team A. Sometimes they play as the visitor. Sometimes they play as the home team. Team B is the Milan Macaroni.

Jerry: Macaroni? That could be offensive.

Jim: Well, leave it as Team B. So who won?[1]

Solution on p. 186

---

[1]Henceforth in this chapter, if one team has the same number of PA in every inning in which it bats, we will say that it played a "steady game."

3.3 A STEADY DOUBLE

Jerry: My turn!

Brooklyn played a steady doubleheader as the visiting team, that is, Brooklyn sent the same number of players to the plate in each inning of the two games. The first was a regular 9-inning game. The second ended in the tenth inning.

Both games had the same score. The only difference (besides the length of the game) is that Brooklyn left 10 on base in the first game and left 13 on base in the second.

How many runs did Brooklyn score in each game?[2]

Solution on p. 187

3.4 CICIONEDDOS

Jim: My turn again.

The Bombshells played the Cerignola Cicioneddos. The 'Shells were the visitors. They scored 2 runs and left 4 on base. Their game was, of course, steady. How many innings were there?

Jerry: Cool! Can I assume they played a regulation game, batting in at least 4 innings?

Jim: Yes.

Solution on p. 187

---

[2]This puzzle first appeared in the authors' "Retrograde Baseball Redux," *The Mathematical Intelligencer* 44(4), 2022.

3.5 MALTAGLIATTI

Jerry: I can't stop thinking about the Bombshells. What a team!

As the home team, Brooklyn played a steady game against the Messina Maltagliatti. They scored only one run and left 13 on base. Mavis Sellary, batting ninth in the order, went 4 for 4 and drove in the run. Who won?

Solution on p. 187

3.6 LUCCA

Jim: You know Jerry, I don't know if these steady games are so great. Most of the problems seem pretty simple. Maybe they're not really as interesting as our puzzles from real box scores, or even our composed problems involving the intricacies of runs and rbis and when players bat.

Jerry: I hear you. So if you want something harder, try this one:

The Bombshells just played the Lucca Luccioli. Both teams played "steady" games (enforced by the Bombshells) but the teams didn't necessarily have the same steady rate. Lucca left 15 runners on base. Brooklyn left 11 runners on base. One team shut out the other. The question is this: Did the winning team score 58 runs?

Jim: That question is totally bizarre!

How I am supposed to figure that out???

Hint on p. 169; solution on p. 188

CHAPTER 4

Secrets of the Past

4.1 ANCIENT RIVALRY

Jerry: Box scores list homers hit during a game. An odd fact is that the homers are listed in the order that they were hit. This fact turns out to be important in analyzing the game at right (Sept. 9, 1962).

The challenge here is to analyze Boston's offense—who scored when, driven in by whom.

Jim: You need that fact about the homers?

Jerry: Actually, I don't know for sure. But I do use it in my solution.[1]

YANKEES

FIRST GAME

BOSTON	ab	r	h	bi	NEW YORK	ab	r	h	bi
Geiger,cf	4	1	0	0	Kubek,ss	3	2	1	0
Bressoud,ss	4	1	1	2	Richardson,2b	4	1	2	0
Yastrzemski,lf	4	3	2	2	Tresh,lf	3	0	2	1
Clinton,rf	5	1	3	2	Maris,cf	3	0	1	2
Runnels,1b	5	1	1	2	Lopez,rf	4	0	0	0
Malzone,3b	4	0	1	1	Howard,c	4	0	1	0
Nixon,c	4	0	1	0	Skowron,1b	4	0	0	0
Schilling,2b	4	1	2	0	Boyer,3b	4	0	0	0
Monb'quette,p	2	1	0	0	Brown,p	1	0	0	0
Total	36	9	11	9	Sheldon,p	0	0	0	0
					aLinz	1	0	0	0
					Cullen,p	0	0	0	0
					bBlanchard	1	0	0	0
					Clevenger,p	0	0	0	0
					cLong	1	0	0	0
					Total	33	3	7	3

aFlied out for Sheldon in the 5th; bFouled out for Cullen in the 7th; cFouled for Clevenger in the 9th.

Boston .... 1 0 0 3 5 0 0 0 0—9
New York .... 2 0 0 0 1 0 0 0 0—3

E—Boyer, Skowron. A—Boston 7, New York 11. DP—Malzone, Runnels; Zipfel; Clevenger, Kubek, Skowron. LOB—Boston 5, New York 6. 2B Hits—Yastrzemski, Richardson, Maris, Tresh. 3B—Malzone, Clinton. HR—Yastrzemski, Bressoud, Clinton. Sacrifices—Monbouquette, Geiger. SF—Bressoud.

	IP.	H.	R.	ER.	BB.	SO.
Monb'q'ette (W, 12—13)	9	7	3	3	3	6
Brown (L, 6—5)	4.2	7	9	4	1	1
Sheldon	0.1	0	0	0	0	0
Cullen	2	1	0	0	1	2
Clevenger	2	3	0	0	0	1

Umpires—Paparella, Soar, Smith, Rice. Time—2.37.

---

[1]This puzzle first appeared in the authors' "Retrograde Baseball Redux," *The Mathematical Intelligencer* 44(4), 2022.

DOI: 10.1201/9781003332602-4

Hint on p. 170; solution on p. 190

4.2 DOUBLE SWITCHES

Jerry: Jim, are you ready for the next level?

Jim: What is this? A video game?

Jerry: No no. I mean the next level of sophistication in analyzing real box scores. Take a look at this, a game played on June 15, 1970, and you'll see what I mean:

MILWAUKEE	ab	r.	h.	bi	BALTIMORE	ab	r.	h.	bi
Harper, 3b	5	2	2	0	Buford, lf	4	1	1	0
Hegan, 1b	4	3	2	2	Salman, 3b	5	2	2	2
Savage, cf	5	1	2	2	F.Robinson, rf	3	1	1	0
Walton, lf	3	0	0	0	Powell, 1b	5	0	0	0
Francona, 1b	1	1	1	1	Rettenm'nd, cf	4	1	2	4
Hershberg'r, rf	3	0	0	0	Johnson, 2b	1	0	0	0
Snyder, lf	1	0	0	0	Floyd, 2b	2	0	0	0
Kubiak, ss	3	1	1	0	May, ph	1	0	0	0
Roof, c	2	0	0	0	Hendricks, c	3	0	1	0
Humphreys, p	0	0	0	0	Richert, p	0	0	0	0
Pena, ss	2	1	1	3	Hall, p	0	0	0	0
Gil, 2b	1	0	0	0	Belanger, ss	4	1	2	0
Alvis, ph	1	0	0	0	Cuellar, p	3	0	1	0
Sanders, p	1	0	0	0	Watt, p	0	0	0	0
Baldwin, p	0	0	0	0	Dalrymple, c	1	0	0	0
Bolin, p	1	0	0	0	Total	36	6	10	6
McNertney, c	3	0	1	1					
Total	36	9	10	9					

```
Milwaukee .......... 2 0 0   0 0 1   0 6 0—9
Baltimore .......... 4 0 0   0 0 2   0 0 0—6
```

E—Johnson, Belanger. LOB—Milwaukee 3, Baltimore 9. 2B—Salman, Pena. HR—Rettenmund (9), Hegan (4), Salman (2). SB—Belanger, Savage, Francona. S—Floyd.

	IP.	H.	R.	ER.	BB.	SO.
Bolin	3.1	6	4	4	2	5
Humphreys	2.2	4	2	2	1	3
Sanders (W, 1—0)	1	0	0	0	0	0
Baldwin	2	0	0	0	1	1
Cuellar	7	7	6	5	2	7
Watt (L, 2—3)	0	1	1	1	0	0
Richert	0.2	2	2	2	1	2
Hall	1.1	0	0	0	0	1

HBP—by Sanders (Rettenmund).
T—2:55. A—6,280.

Jim: That's crazy! There are 17 players listed just for one team! And look, the fielding positions are all mixed up. I thought you couldn't do that! The first basemen, shortstops, and pitchers are scattered around in multiple places in the lineup!

Jerry: Let's do Baltimore. They only have 15 players.

Jim: Thank you.

Jerry: The key is to look at the pitching stats at the bottom of the box score. In the pitching section, box scores list each team's pitchers in the order in which they pitch. The first Baltimore pitcher listed is Cuellar, followed by Watt, who lost the game. So Cuellar pitched first. The 7 in the IP column means he pitched 7 full innings and possibly pitched into the eighth without getting any outs.[2]

Jim: Hey! After Watt comes Richert, but in the batter stats he's listed way up in the lineup, like he replaced the catcher, Hendricks!

Jerry: And that's what must have happened. In the eighth inning, when Richert replaced Watts as pitcher, Baltimore also brought in a new catcher, Dalrymple, to replace Hendricks. When two substitutions are made at the same time, the manager is allowed to choose in which open slot they will bat. Baltimore chose to put the pitcher in the catcher's slot and vice versa, a so-called double switch.[3]

Jim: Whew. Okay. Then after that, Richert pitches for 2/3 of an inning and is replaced by Hall, still in the eighth inning. And that's not a double switch, right? Hall bats in the same slot as Richert.

---

[2]Some box scores note when a pitcher pitches a part of an inning without getting any outs by including a footnote to that effect. But, as we shall see, this box score doesn't follow that convention.

[3]Since the pitcher and catcher switched batting slots, we can conclude logically that it must have been a double switch, with both substitutions occurring at the same time. We don't have to know why the manager made that choice. But if you're curious, it's usually because the manager wants to have his better hitters batting more often. Pitchers are often poor hitters, and get switched into a batting slot that won't come up soon.

Jerry: Right. Is that all for Baltimore's switching?

Jim: Well yes, Hall is the last pitcher. Oh wait. Something happened at second base. I guess Floyd replaced Johnson for some reason, and then May pinch-hit for Floyd. That must have been early—Johnson only got 1 at-bat.

Jerry: Here's the order of the switches in your scenario:

		eighth	eighth	eighth	
Buford					
Salmon					
Robinson					
Powell					
Rettenmund					
Johnson	Floyd				May
Hendricks			Richert	Hall	
Belanger					
Cuellar		Watt	Dalrymple		

Jerry: But when did Floyd pinch-hit for Johnson? And when did May pinch-hit for Floyd?

Jim: I don't know. I'm new at this!

Jerry: And besides, how do you know that your scenario is the right one? Maybe Johnson substituted for Rettenmund in slot 5 instead. In that case, we'd have Floyd as the starting second baseman in slot 6. Can you rule out that possibility?

Jim: I'm still new at this.

Jerry: What can you say about Baltimore's scoring?

Jim: Everything! I'm an expert.

Solution on p. 194

4.3 SWITCH CENTRAL

Jerry: Okay, that was the easy part.

Jim: Easy?

Jerry: Now it's time to track Milwaukee's pitching and scoring.

MILWAUKEE	ab	r.	h.	bi	BALTIMORE	ab	r.	h.	bi
Harper, 3b	5	2	2	0	Buford, lf	4	1	1	0
Hegan, 1b	4	3	2	2	Salman, 3b	5	2	2	2
Savage, cf	5	1	2	2	F.Robinson, rf	3	1	1	0
Walton, lf	3	0	0	0	Powell, 1b	5	0	0	0
Francona, 1b	1	1	1	1	Rettenm'nd, cf	4	1	2	4
Hershberg'r, rf	3	0	0	0	Johnson, 2b	1	0	0	0
Snyder, lf	1	0	0	0	Floyd, 2b	2	0	0	0
Kubiak, ss	3	1	1	0	May, ph	1	0	0	0
Roof, c	2	0	0	0	Hendricks, c	3	0	1	0
Humphreys, p	0	0	0	0	Richert, p	0	0	0	0
Pena, ss	2	1	1	3	Hall, p	0	0	0	0
Gil, 2b	1	0	0	0	Belanger, ss	4	1	2	0
Alvis, ph	1	0	0	0	Cuellar, p	3	0	1	0
Sanders, p	1	0	0	0	Watt, p	0	0	0	0
Baldwin, p	0	0	0	0	Dalrymple, c	1	0	0	0
Bolin, p	1	0	0	0	Total	36	6	10	6
McNertney, c	3	0	1	1					
Total	36	9	10	9					

Milwaukee .... 2 0 0   0 0 1   0 6 0—9

Baltimore .... 4 0 0   0 0 2   0 0 0—6

E—Johnson, Belanger. LOB—Milwaukee 3, Baltimore 9. 2B—Salman, Pena. HR—Rettenmund (9), Hegan (4), Salman (2). SB—Belanger, Savage, Francona. S—Floyd.

	IP.	H.	R.	ER.	BB.	SO.
Bolin	3.1	6	4	4	2	5
Humphreys	2.2	4	2	2	1	3
Sanders (W, 1—0)	1	0	0	0	0	0
Baldwin	2	0	0	0	1	1
Cuellar	7	7	6	5	2	7
Watt (L, 2—3)	0	1	1	1	0	0
Richert	0.2	2	2	2	1	2
Hall	1.1	0	0	0	0	1

HBP—by Sanders (Rettenmund).
T—2:55. A—6,280

Jim: Well, Bolin is the first pitcher. He pitches 3 1/3 innings and in the middle of the inning—ooh it's a double switch—he's replaced by ... Humphreys, who will bat in the catcher's slot, while McNertney comes in to catch and bat in slot 9.

Humphreys lasts 2 2/3 innings and is replaced by Sanders, who ... will bat in the second baseman's slot! It's another double switch! Wait. There's this guy, Alvis who pinch-hits for the second baseman, Gil. Well then, who will play second base? Help!

Jerry: I thought there might be problems. Let's see if we can get the slots right. Milwaukee has 27 outs, 9 runs, and 3 LOB for 39 PA. That means the first 3 slots have 5 PA and all the rest have 4. Milwaukee batters have 36 ab and Baltimore pitchers Cuellar and Richert give up 3 walks, so that accounts for all 39 PA, so … Hegan must get a walk so he'll have 5 PA. The next slot has Walton with 3. I think Francona replaces Walton, filling out the 4 PA of slot 4.

Jim: But maybe Walton walked once. Maybe instead, Francona amd Hershberger share slot 5!

Jerry: Hmm … but whomever Francona is paired with, he can have only 1 of the 4 PA in his slot; if he starts the game in slot 5, he would have to score in the first inning. But he can't by the Gap Rule (p. 24).

Jim: Okay. And besides, if Francona started the game, we'd have two first baseman—that can't be right either.

So Hershberger is in slot 5, and Kubiak in slot 6. But what about Snyder—is he in slot 5 after Hershberger, or in slot 6 before Kubiak? Oh, he can't be in slot 6, or there would be 2 starting left fielders, so he must be in slot 5. So Kubiak must have a walk for his fourth PA. Roof, Humphreys, and Peña share slot 7. Gil, Alvis, and Sanders share slot 8, and Baldwin, Bolin, and McNertney share slot 9. But how and when does this all happen, with all positions always filled?

Jerry: The Bolin/McNertney/Humphreys/Roof switch is in the middle of the fourth, as Bolin pitches 3 2/3 innings. You lose a pitcher and a catcher and get a pitcher and a catcher.

Jim: Ah, then Humphreys completes the sixth inning. When he leaves, Peña takes his place in the lineup, Sanders comes in and takes the place of Gil/Alvis. That means that Alvis must pinch-hit for Gil in the bottom of the sixth. So this double (triple?) switch takes place before the seventh. Hey! You lose a second baseman (Gil) and a pitcher and get a pitcher and a shortstop (Peña)?

Jerry: That's right, but you can't have 2 shortstops at the same time, and you always have a second baseman. So the previous shortstop, Kubiak, must have moved to second base to take the place of Gil. Ideally, the box score should say "Kubiak, ss-2b," to make this clear. But sometimes they don't bother to record changes in position.

Jim: At some point, Snyder replaces Hershberger. You lose an outfielder and gain an outfielder. No problem, I guess.

Sanders pitches the seventh and Baldwin replaces him to pitch the eighth and ninth. That's all the switches involving pitchers.

Jerry: We have a couple more substitutions that don't involve the pitchers at all. Francona comes in at first base, replacing the left fielder, Walton. But there's no replacement for the starting first baseman, Hegan. We can't have two first basemen at the same time, Hegan and Francoma, can we?

Jim: Certainly not. The simplest resolution is that Hegan moves to right field to take over Hershberger's original position.

Jerry: Okay then, now for the scoring. Maybe this will be easy ...

Hint on p. 170; solution on p. 195

4.4   A WORLD SERIES GAME

Jerry: Jim, Nancy sent me a box score. She's been doing that lately. Either she's getting interested in the puzzles or ...

Jim: Or ... ?

Jerry: Or she's just needling me. I can't tell.

Jim: Why would she do that?

Jerry: Well she's my younger sister. So as a kid mostly I ignored her. Or if we had a dispute, I was bigger and stronger, so I tended to get my way.

In order to get even, she had to be more subtle. Maybe she would "accidentally" knock over my toy soldiers. Or she'd find some way to tease me until I got really mad, and then of course my parents would blame me!

Jim: But I thought you got along great with her!

Jerry: Yes, we do *now*. But she still likes to tease me. In a good-natured way, of course.

Jim: Of course. If she's really forgiven your childhood sins . . .

Jerry: Back to the box score.

Nancy picked a good one. Take a look. It's the fourth game of the 1982 World Series. The Brewers facing the Cardinals.

Brewers 7, Cardinals 5

GAME FOUR

ST LOUIS	ab	r.	h.	bi	MILWAUKEE	ab	r.	h.	bi
Herr 2b	4	0	0	2	Molitor 3b	4	1	0	0
Oberkfell 3b	2	2	1	0	Yount ss	4	1	2	2
Tenace ph	1	0	0	0	Cooper 1b	4	1	2	1
Hrnandz 1b	4	0	0	0	Simmons c	2	0	0	0
Hendrick rf	4	0	1	1	Thomas cf	4	0	1	2
Porter c	3	0	1	0	Oglivie lf	3	1	1	0
L.Smith lf	4	1	1	0	Money dh	4	2	2	0
Iorg dh	4	0	2	1	Moore rf	4	0	1	0
Green pr	0	0	0	0	Gantner 2b	4	1	1	1
McGee cf	4	1	1	0					
O.Smith ss	3	1	1	0					
Total	33	5	8	4	Total	33	7	10	4

St. Louis ....1 3 0   0 0 1   0 0 0—5
Milwaukee ....0 0 0   0 1 0   6 0 x—7

E—Gantner, Yount, LaPoint. DP—St. Louis 2, Milwaukee 2. LOB—St. Louis 6, Milwaukee 6. 2B—Oberkfell, Money, L. Smith, Iorg, Gantner. 3B—Oglivie. SB—McGee, Oberkfell. SF—Herr.

	IP.	H.	R.	ER.	BB.	SO.
St. Louis						
LaPoint	6.2	7	4	1	1	3
Bair L,0-1	0	1	2	2	1	0
Kaat	0	1	1	1	1	0
Lahti	1.1	1	0	0	1	0
Milwaukee						
Haas	5.1	7	5	4	2	3
Slaton W,1-0	2	1	0	0	2	1
McClere S,1	1.2	0	0	0	0	2

WP—Haas, Kaat. T—3:04. A—56,560

The box score looks extremely challenging. I'm going to need your help.

Jim: Well. Seems okay I guess. No double switches. Lots of runs.
. . .

Jerry: Look at the bottom lines on pitching.

Jim: Hmm… Ah! St. Louis has more runs than earned runs. Hey! So does Milwaukee!

Jerry: Now check the bottom lines on hitting.

Jim: Oh yeah. St. Louis scored more runs than they drove in. Oh good grief. So did Milwaukee!

Jerry: Finally, I'm curious about Herr.

Jim: He leads off for the Cards. He has 2 runs batted in. Oh I see what you mean. He has no hits! How does he drive guys in?

Jerry: He does have a sacrifice fly. That accounts for one of the rbis. And he could have had an rbi on a ground out, or a bases-loaded walk, or something.

Jim: Okay. I agree. This is a challenge.

Jerry: Because of Herr, let's tackle the St. Louis side first.

Jim: Okay, well what I like to do is start with the runs—figure out who drives in which runs.

Jerry: That's good for most box scores, but in this game not every run *was* driven in. One wasn't and we don't know which that is. I suggest that instead of looking at runs, we look at the rbis. After all, every rbi does drive in a run.

Jim: Good idea. So let's start with your favorite player, Herr. He leads off the first, and doesn't have a homer, so he can't get an rbi in the first inning. The only other times he can have an rbi are the second and sixth innings.

He has to have at least 1 rbi in the second inning because only one run scores in the sixth. Could he bat in 2 runs in the second? He'd have to have 2 at-bats in the second, one for a sac fly, one for a ground out or something else. But that's impossible; it would mean at least 10 PA in the inning, which would mean at least 4 runs, but St. Louis scores only 3 runs in an inning.

So Herr must drive in 1 run each in the second and sixth innings, the only other innings in which St. Louis scores.

Jerry: OK, we're getting somewhere. But look—in the sixth inning Herr can't drive in Oberkfell, because he bats before Oberkfell. He must drive in somebody else for the inning's only run. So Oberkfell can't score in the sixth inning. He must score in the first and second innings if he's going to get his 2 runs.

Jim: Right. So now let's look at Iorg's rbi. When does he get his rbi? Not in the sixth, because Herr has that rbi.

Jerry: Good. And Iorg can't drive in Oberkfell in the first inning. Iorg is the seventh batter in the inning. After 6 batters there would be at most 2 outs and 3 on base. Somebody has to have scored, so it has be Oberkfell, the only run of the inning. So it wasn't Iorg who drove him in.

Jim: So Iorg has to get his rbi in the second inning. Hey, this is easy! And who does he drive in? Not Oberkfell, who will bat after Iorg in the second inning. The only other batter hitting before Iorg in the second inning with a run is L. Smith, so that's who he drove in.

Jerry: OK, so now we've got Oberkfell and L. Smith scoring in the second inning, and we need a third run, but no more. So one of McGee and O. Smith score—the other scores in the sixth inning.

Jim: Wait a minute. Between L.Smith and Oberkfell, who score the first and last runs of the inning, we've got four players—Iorg, McGee, O.Smith, and Herr—who must score or be out. Only one scores, so the other three must be out. And they must be out before Oberkfell can score. Wouldn't that end the inning with only 2 runs?

Isn't that a contradiction?

Jerry: **?**

Jim: **?**

Jerry: **?**

⟨Five minutes pass⟩

Jerry: It looks like we made a mistake here. But I don't see any error in our logic.

Jim: Every step was good. It was logic. All of it.

Jerry: Unless we made some unjustified assumption.

Jim: But what assumption? Where did we start? We were looking at Herr...

Jerry: His 2 rbis, his sacrifice fly ...

Jim: We said that he had one in the second and one in the sixth.

Jerry: But that wasn't an assumption. That followed from the fact that he couldn't drive in 2 runs in the second.

Jim: But *that* wasn't an assumption. That followed from the fact that he needed 2 at-bats to drive in 2 runs.

Jerry: But *that* wasn't an assumption. That followed from the fact that you can't drive in two guys with a sacrifice fly.

Jim: But *that* wasn't–

Jerry: Wait! That's the assumption. **That's the assumption!**

Jim: Two runs on a sacrifice fly? Is that possible?

Jerry: It could happen if we were making up a problem to put in a book. But this is real life. This game actually happened.

Jim: Crazy things happen. I read that somewhere.

Jerry: Here's what we've just done:

We assumed Herr couldn't drive in two with sac fly.
That led us logically to a contradiction.
That's a *proof* that he did.

We proved that Herr drove in two guys with a sac fly.

⟨pause⟩

Jim: Nancy picked a winner, didn't she?

Jerry: Yeah.

Do you suppose she knew what happened in that game and wondered if we could prove it?

> Jim and Jerry had to check their logic several more times before they were sure about the 2-run sacrifice fly. Almost everything they thought they had proved was tainted by their false assumption.
>
> Who *really* scored the runs in each inning? You'll have to start over to figure that out. But you have a head start. You know about the 2-run sacrifice fly!

Solution on p. 197

4.5   THE OTHER SIDE OF THE WORLD

Jerry: The Brewers' offense in that fourth game of the World Series presents interesting problems too.

Jim: Their seventh inning looks especially tricky.

For the Brewers, I think the key is the pitching record. Maybe we can figure out what each seventh inning pitcher did.

Jerry: Good. And we should keep in mind that 6 runs have to be scored before 3 outs. Every player between the first and last scorer must either score or be retired.

Solution on p. 198

4.6 LEARY'S BAD DAY

Angels 9, Yankees 5

YANKEES	ab	r	h	bi	CALIFORNIA	ab	r	h	bi
Kelly cf	5	2	4	1	Polonia lf	4	2	3	4
Sax 2b	5	1	1	0	DHill 2b	3	1	1	0
Azocar lf	5	0	1	0	CDavis dh	3	1	1	3
Balboni dh	4	0	2	2	Winfield rf	4	0	0	0
Hall ph	1	0	0	0	Stevens 1b	4	1	0	0
Maas 1b	3	0	1	1	Parrish c	2	2	0	0
JeBrfld rf	3	0	0	0	DWhite	3	1	1	2
Leyritz 3b	4	1	1	1	KAnders 3b	3	0	0	0
Geren c	3	1	1	0	Schofield ss	2	1	0	0
Nokes c	1	0	0	0					
Espnoz ss	4	0	1	0					
Total	38	5	12	5	Total	28	9	6	9

Yankees ....1 0 2 1 0 1 0 0 0—5
California ....0 4 0 0 5 0 0 0 x—9

E—KAnderson, Leary. DP—Yankees 1. LOB—Yankees 8, California 3. 2B Hits—Kelly, Maas, Polonia. HR—Polonia (2), Leyritz (4), CDavis (12), DWhite (9). SB—Kelly (26). S—DWhite, DHill.

	IP	H	R	ER	BB	SO
Yankees						
Leary L, 6-15	4.2	5	9	6	4	7
JDRobnsn	1.1	1	0	0	1	1
Plunk	2	6	0	0	0	4
California						
Abbott W, 8—10	5.2	11	5	5	0	2
Fraser S, 1	3.1	1	0	0	2	0

HBP—Parrish by Leary. WP—Leary.
Umpires—McCoy, Phillips, Cousins, Hirschbeck.
T—2.37. A—27,937

Jerry: What a day for Leary![4] The Angels hit him for 9 runs and 3 homers in 4-plus innings. He hit a batter. He threw a wild pitch. And he committed the only Yankee error![5]

I (naturally) tried to answer six questions:

1. In what inning did Polonia hit his homer?
2. In what inning did Davis hit his homer?
3. In what inning did White hit his homer?
4. In what inning did Leary make an error?
5. In what inning did Leary hit Parrish?
6. In what inning did Leary make a wild pitch?[5]

I got answers to four of them. Then I proved that two of them can't be solved. Your turn!

Jim: You're really piling it on.

Jerry: I can't help it.

Hint on p. 171; solution on p. 201

---

[4] August 14, 1990

[5] Just a reminder: A wild pitch isn't counted as an error since it was not committed by the player in his role as a fielder. Leary's error, wild pitch, and hit batsman are three separate incidents.

CHAPTER 5

Puzzles as Art

5.1 NEVER OUT, NEVER IN

Jim: Here's what I call an artistic puzzle. It's a mystery. You're given hardly any clues. But somehow, somehow it's possible to solve the mystery:

There's a guy playing for a team who had a special day. He was never out and he was never in—"never in" meaning he never scored. All the same, he led his team to a 6–5 victory when he hit a two-out walk-off triple in his fourth plate appearance.

Who was he, that is, what was his place in the batting order?

Say, Jerry, why don't you make up an artistic puzzle?

Jerry: I'm working on one. But it's a different kind of artistic puzzle. It has lots of clues, but it's hard to tell what they really mean.

Jim: That sounds like you. I often have no idea what *you* really mean!

Solution on p. 205

DOI: 10.1201/9781003332602-5

5.2 DOUBLE PLAYS

Jim: Here's another beautiful puzzle (I'm saying so myself):

Foster, playing for the Birmingham Blackbirds, hit into double plays every time he batted. More than once, though, a single run scored when he hit into a double play. The Birds didn't score any runs in any other way. They left nobody on base. Somehow, though, the Birds managed to pull it out, beating the Tallahasse Tunafish in extra innings. What was the final score?

Jerry: Hmm ... A run can score on a double play, but there would have to be at least two players on base, one to be out on the double play, and one to score. He could be the third batter up in the inning, but he couldn't be one of the first two.

Jim: Could the Birds score a run when Foster was the fourth batter up in an inning?

Jerry: Let's see—then he'd have the last plate appearance in the inning, because there are no LOB and only 1 run, and 3 outs—just 4 PA. But he could bat with the bases loaded, hit into a double play on which a run scores, and then be picked off base. So that's possible, too.

Jim: That's good.

Jerry: So Foster can be the third guy up in an inning. Or the fourth.

Hint on p. 171; solution on p. 206

5.3 MISTAKES WERE MADE

Jerry: Those are great puzzles, Jim! But great puzzles come in different shapes and sizes. I'm attracted to puzzles that are intricate and involved, that take the solver down a long and winding path, with surprising turns and new ideas.

They're difficult to construct! But I've constructed one for your enjoyment. It's a game in which three players, Blanco, Gato, and Herrera, committed errors.

Cats	ab	r	h	bi	Dogs	ab	r	h	bi
Abe ss	4	1	0	0	Amsler 1b	4	1	1	0
Brown 3b	3	1	2	1	Blanco cf	3	1	1	1
Castro cf	4	0	0	0	Clark lf	4	0	0	0
Dominguez c	4	1	0	0	Durand rf	4	1	1	0
Engel rf	4	1	1	1	Emerson 2b	4	0	1	1
Frank 1b	3	0	1	0	Foster 3b	3	0	0	1
Guzman lf	3	0	0	0	Gato c	3	0	0	0
Hall 2b	4	0	0	1	Herrera ss	4	0	1	0
Iglesias p	2	0	0	0	Ito p	2	0	0	0
Ivory ph	1	0	1	0	Ignacio p	1	0	0	0
Ingals p	1	0	0	0	Ichikawa p	1	0	0	0
Totals	33	4	5	3	Totals	33	3	5	3

```
Cats   2 0 0   0 0 0   1 0 1--4
Dogs   0 0 0   0 0 0   0 0 3--3
```

		ip	h	r	er	bb	so
Iglesias	(W)	7	2	0	0	1	5
Ingals		2	3	3	3	2	2
Ito		6	2	2	1	0	1
Ignacio		2	3	1	0	1	0
Ichikawa	(L)	1	0	1	0	2	2

```
LOB--Cats 5, Dogs 6, HR--Brown, Engel,
Foster, 2b--Frank, E--Blanco,
Gato, Herrera, GIDP--Ingals
```

Jerry: When was each error committed and under what circumstances?

And exactly what happened in the Cats' ninth inning?

Jim: Ooh.

Ack!

The centerfielder, the shortstop, the catcher.

Oof.

Hint on p. 171; solution on p. 206

5.4 THE TWO RODS

Jim: That last puzzle was amazing. I definitely needed the hints.

But here's another puzzle that I hope has elegance:

Anytown played a game with Otherburg. It was a lovely sunny day, but one guy on Anytown, Gus Rodriguez (GRod), batting in slot 6 of the lineup, had a terrible game. He was up four times. He went down swinging in each of his at-bats ending the inning for Anytown; indeed, the last plate appearance in the game for Anytown was his. Worse, he twice struck out when there were 2 outs and the bases were loaded!

Oddly, Otherburg had a guy with the exact same story. The only difference was that he batted in slot 7 of Otherburg's lineup. Curiously, his name was also Rodriguez, Hernando Rodriguez (HRod).

Both teams scored. But which team won?

Jerry: I do like this. It has symmetry. And it has asymmetry.

Solution on p. 210

5.5 THE GUY WHO DID IT ALL

Jim: This is an old puzzle.[1] It has features of both sorts of puzzlish beauty. It's short, just a few clues. But it takes the solver down a new path:

The Blue Sox was the home town team. They won the game 6–4. One of their players, Ike Farrell, scored all 6 runs.

How many players did the Blues leave on base in the seventh inning?

Jerry: The last sentence hits you with a thud. What a bizarre question! How could we possibly know the answer?

Except, of course, we know the answer.

Solution on p. 211

---

[1]The puzzle first appeared in the authors' "Baseball Retrograde Analysis," *The Mathematical Intelligencer* 40(4): 71-76, 2018.

5.6 SCRAPS

Blanco cf 5 1 0 0

8 11 8 39

Totals

Jim: Jerry, I got what I think is a puzzle in the mail. It's a lot of pieces of paper.

Eastham

Padbourn o 4 0 1 O

h r er BB SO

3

Kimura

—6—

2

0

0

0

0

0

0

4

Westwood

O

di

Jackson

O

O 1 1 e 3

6 2/3 9 4 4 O

Shugart 1b

aв r h

Duffy 1b

Staatz w

3 0 0 0 0 0 0 0 c 3

1 2 2 1

4 1 2

1.1/3 1 2 2

5 1 0 1 2

Smith 2B 5 1 2 0

Namath 3B 5 1

Okada ss

Suarez rf 5 1 2 3

4 1 1 2

Martin P 4 1 1 O

2—8

LOB—Eastham 9, Westwood O. HR—Okada, Namath. HBP—Duffy by Faatz.

Jerry: Who sent it?

Jim: There's no return address.

Jerry: Mysterious!

Jim: Looks like it's pieces of a box score. It's some sort of a puzzle. We're supposed to deduce something about the game from this.

Hey, could it be ... There's only one person who knows we're doing baseball puzzles, your sister Nancy.

Maybe she ...

Jerry: Couldn't be Nancy.

Jim: Why not?

Jerry: Couldn't.

So you'd better get to work.

Jim: What?

Me?

Not *us?*

Jerry: You.

Jim: Well, ...

⟨pause⟩

Well, ...

⟨pause⟩

Well, okay.

⟨pause⟩

I count 9 different players labeled with their fielding positions and 4 numbers after them. That sure looks like the batting stats (ab, r, h, bi) for a team.

And we have a line of totals: 39, 8, 11, and 8. It looks like the team scored 8 runs on 11 hits. I'll bet if you added up the numbers on the 9 lines it would match.

Jerry: Looks like it does.

Jim: And I'll bet that team is Eastham, because I see a line score in which Eastham has 8 runs. There's another line score for Westham, which has 6 runs. Eastham won 8–6.

Hmm. There are some other lines with six numbers in them, and the first one is sometimes a fraction! Oh, it must be the pitching lines, and the fractions must be partial innings pitched. But which team are they on? ... Hmmm. The pitcher in Eastham's batting lines is Martin, and he's not in the pitching lines at all, so it can't be Eastham. It must be the pitching lines for Westwood.

Then we have the batters from Eastham and the pitchers from Westwood. And Westwood pitched only 8 innings (6 1/3 + 1 2/3). From the line scores, Eastham won in a walk-off, scoring 2 runs in the ninth. Eastham must have been the home team.

Jerry: Ah. That's helpful.

Jim: Okay, puzzle solved!

Jerry: What? Aren't you going to arrange those pieces of paper together into a box score?

Jim: No, that's ridiculous. There's no way anybody could do that.

Jerry: How do you know if you haven't even tried?

Jim: And how do you know what the mystery person had in mind sending me this puzzle?

Jerry ...

Jerry, did you send me this puzzle?

Jerry: Why don't you just give it a try?

Jim: Wait. Just a minute. . . . I'm arranging the players . . .

⟨A minute⟩

Ah. There.

	aB	r	h	Bi
Blanco cf	5	1	0	0
Duffy 1B	4	1	2	1
Martin p	4	1	1	0
Namath 3B	5	1	1	2
Okada ss	4	1	1	3
RadBourn c	4	0	1	0
ShuGart lf	3	1	1	0
Smith 2B	5	1	2	0
Suarez rf	5	1	2	2

	ip	h	r	er	BB	so
Kimura	6 2/3	9	4	4	0	3
Jackson	0	—	2	2	1	0
Faatz	1 1/3	—	2	2	0	1

Jerry: That's not right. That's a mess. How could you think—

Jim: Give me a break! I'm just putting them in alphabetical order. Eastham batters on the left, Westwood pitchers on the right.

Jerry: Oh.

Jim: And here's the other stuff:

Westwood	4	0	0	0	0	0	0	0	2	—6
Eastham	3	0	0	0	0	0	3	0	2	—8

L OB—Eastham 9, Westwood 0, HR—Okada, Namath, HBP—Duffy By Faatz

Jim: Okay now, from LOB = 9, runs = 8, and outs = 24, Eastham
        had 41 PA. That means the first 5 batters had 5 PA each, and
        the other four had 4. Namath, Suarez, Blanco, and Smith
        each had 5 AB, and Duffy had 4 AB plus a HBP. Thus they,
        in some order, were the first 5 batters.

        Let's check the at-bats: from the totals line, there were 39
        AB. There's also one HBP, and we're going to need one walk
        to reach 41 PA. That's the walk in Jackson's pitching line.
        He must have walked Shugart, who only had 3 AB. OK,
        that's in agreement.

        The "walk-off" was Eastham's 2 runs in the ninth. The batter
        in the fifth slot drove in the winning run. That could have
        happened in lots of different ways.

Jerry: Actually, only one way.

Jim: One way? But there's lots of people scoring, lots of people
        driving people in! Anybody with an rbi could be the guy in
        the fifth slot.

Jerry: I mean, could Eastham have scored the 2 runs separately?

Jim: Jerry, did you send this?

Jerry: Could Eastham have scored the 2 runs separately?

Jim: Sure. Get one run, then get ... Oh wait. Eastham needed
        only one run to win. If they got one run first, the game would
        be over. Okay, so the walk-off hit drove in 2 runs.

        Ah! We know the first 5 batters. Two of them have 2
        rbis—Namath and Suarez. Either one could have hit the
        winner.

Jerry: Actually, no.

Jim: Oh oh oh. Normally in a walk-off, only the winning run is
        recorded, but there's an exception in the case of a home run.
        With a homer, all the runs count. I'll bet that's because the
        runs all happen at the same time (when the ball goes over the
        fence).

        Okay, so then the fifth batter has to be Namath, the only one
        of the batters with 5 plate appearances who hit a home run.
        Got it.

Say, I think this means there were just 2 batters in the ninth.

Jerry: Why?

Jim: There were no outs in the ninth.

Jerry: Good! Why?

Jim: Um um Westwood pitched just 8 innings!

Okay there were 2 runs and no outs. And no one left on base because of the homer!

Let's see. Who was pitching for Westwood in the ninth? Could it be Jackson? He gave up 2 runs—just what they scored in the ninth. And he got nobody out. I'll bet it was him. He pitched to the batters in slots 4 and 5, and they both scored. Game over.

Jerry: Did Jackson pitch to anyone in the eighth?

Jim: Of course not, he would have racked up an out if he had, but he has no outs.

Jerry: Which out in the eighth?

Jim: Oh, the last out. He'd have to finish the inning, get an out, to go on to the ninth. So he didn't pitch in the eighth.

Jerry: Then how do you account for that walk that Jackson gave up to Shugart?

Jim: Hmmm … Shugart had only 4 PA. He couldn't be in any of the first 5 slots. So he's not the other guy who scored in the ninth. Oh, Okay. Jackson didn't pitch the ninth.

Jerry: Then in the ninth …

Jim: Well, Kimura didn't pitch in the ninth. If he did then he also pitched the sixth, giving up 3 runs there plus 2 in the ninth. But he only gave up 4!

So Faatz pitched the ninth. And in the ninth Faatz pitched to just 2 batters, batters #4 and #5 in the order—if an earlier batter had come up he would either be out or driven in by Namath, but only 2 runs were scored and Faatz didn't record any outs in the ninth.

But wait! How did batter #4 get on base? Faatz gave up only one hit, the home run. Faatz didn't walk anybody, and there were no errors in the game!

Jerry: Don't forget, you can get on base if you're hit by a pitch.

Jim: Oh right! And Duffy was hit by Faatz. That had to happen in the ninth, so Duffy is in slot 4.

⟨Pause.⟩

Jim: Jerry, your puzzle,

assuming I get to the end of it,

is a

Masterpiece.

Solution on p. 212

CHAPTER 6

Baseball Archaeology

6.1 AT THE DAWN OF TIME

Jim: Jerry, have you ever wondered what early box scores looked like?

Jerry: No.

Jim: Well, I did. So I searched the web, and here's what I found: the record of a game between Cartwright's team and Curry's team. It's dated October 6, 1845.

	HANDS OUT			**RUNS**
CARTWRIGHT'S TEAM				
	1	**2**	**3**	
Cartwright	o1	o2		x
Moncrief	o2	o3		x
De Witt	o3			xx
Tucker				xxx
Smith		o1	o1	
Birney				
Brodhead			o2/o3	x
Total				**8**

DOI: 10.1201/9781003332602-6

HANDS OUT	RUNS

CURRY'S TEAM

	1	2	3	
Curry	o1	o2/o3		xx
Niebuhr			o1	xxx
Maltby			o3	x
Dupignac	o2			xx
Turney		o1		xx
Clare			o2	
Gourlie	o3			x
Total				**11**

Umpire - Wheaton

[Data from http://boxscorecards.com/baseball-1800-series.]

What do you think?

Jerry: This is *really* different! It looks like there are statistics for only two things: outs and runs. And for some reason the outs are called "hands out." The x's must be a tally of how many runs each player got, since there are 8 of them for Cartwright's team and it matches the total at the bottom.

The o1, o2, and o3 must be outs, and since there are 3 columns which each have o1, o2, and o3, it must be that they are the 3 outs in each inning. So this box score tells who made each of the 3 outs in each inning. But the whole game had only 3 innings, and there were only 7 players on each team!

Jim: That's what I was thinking. But two questions bother me. Question one is, what's the deal with the "o2/o3"? Is that a double play, or is it two separate outs with a lot of batters in-between?

Jerry: Hah. Hmm … I don't know. *Yet.*

Jim: And question two is, who gets out? For example, on Curry's team in the first inning, it looks like Curry is out on the first play. Then I guess Niebuhr and Maltby get on safely.

And then Dupignac comes up and makes an out. But maybe he hits the ball and Cartwright's team chooses to tag out Niebuhr instead of Dupignac. You know, a fielder's choice. Can "o2" mean that?

Jerry: Yeah. That seems possible. But really we don't know. Yet, I mean.

Jim: Um ... I guess we don't know ... yet.

The website where I got this says that at that time, games were usually played until one team's score reached 21. I get the impression that most of the time players got on base safely, that outs were less common.

Jim and Jerry started by tackling the record for Curry's team. Particular problems for you:

1. Using the box score for Curry's team, determine what it means for an out to be listed next to a particular player. Does it mean (A) that the out occurred during that player's plate appearance, but perhaps a different player was put out, or (B) that this particular player was put out himself, either when he batted, or perhaps later, when when he was a runner?

2. What does the o2/o3 for Curry mean?

3. Who scored when for Curry's team?

Hint on p. 172; solution on p. 214

6.2 LATER THAT DAY

> Jim and Jerry later tackled the box score of
> Cartwright's team. They were able track down all the
> runs. They were helped by the website, which told
> them that in the first inning only three men came up
> and each made an out.
>
> Perhaps you can track down the runs, too.

Solution on p. 217

6.3 ANCIENT HISTORY

Jim: Box scores were quite different a hundred years ago. What do you make of this, from the *New York Times* of July 23, 1923?

It doesn't tell you who had rbi's. Instead, it tells you about their fielding—the number of putouts (Po) and assists (A) they made.[1]

Jerry: This is cool! I wonder what we can deduce from this information.

Well, let's see. Washington had 3 runs, one each in the first, third, and seventh innings. And scoring were Leibhold, Rice, and Evans . . .

•

•

•

Jerry: Can we do anything with the putouts and assists?

Jim: Well this is odd. The catcher has the most putouts. Ruel has eight!

Jerry: I'll bet he gets credit for a putout every time the pitcher strikes out a batter.

Jim: Why is that? All he does is catch the ball.

Jerry: But he *has* to catch the ball. If he doesn't catch it, the batter isn't out.

Jim: Oh, right! But if he drops the third strike, the batter takes off for first, and the catcher throws to first in time to get him out, the catcher could get an assist for that, right? And the first baseman would get a putout.

---

[1]There's a "putout" for every out. A putout is generally credited to the player holding the ball when the out is made. An assist is credited to any player helping to move the ball along. For example, a grounder is hit to the third baseman who throws it to first for the out. The third baseman gets an assist and the first baseman gets the putout. Or say the right fielder catches a deep fly and then throws the ball to the shortstop who relays it to the third baseman in time to tag out a man trying to advance from second to third. The right fielder and the third baseman each get a putout; the right fielder and the shortstop both get assists.

Jerry: Anyhow, this fits. Johnson struck out five Cleveland Indians so five of Ruel's eight putouts are from strikeouts. The others could be from pop-ups. Or guys getting tagged by Ruel when trying to score.

Jim: And the Cleveland pitchers have a total of just 2 strikeouts. That's why the Cleveland catchers have fewer putouts.

Jerry: Hey look at this: Edwards, the Indians' second pitcher has a strikeout. It's in the ninth inning, the only inning he pitched. But the substitute catcher, Myatt, has no putouts.

Jim: Do we know for sure that Myatt was the catcher for that strikeout? The box score doesn't say when he came in. Cleveland is the home team, batting in the bottom half of every inning, so maybe Myatt pinch-hit for O'Neill in the bottom half of the ninth inning.

Jerry: But then why does the box score list him as a catcher, if he never played in the field?

Jim: All right, then maybe he substituted for O'Neill after 1 or 2 outs in the top of the ninth, and Edwards got his strikeout while O'Neill was still catching. I'm still not convinced there's a problem here.

Can you straighten this out? Is it possible to answer the questions below?

Who was catching when Edwards struck out a batter?
Was a third strike dropped in the ninth?
If so, was the runner out at first?
And by the way, can you locate when Rice hit his homer?

The solution is on p. 220

6.4   BASEBALL AT WAR

Jerry: Here's a game from 1944 when many players weren't on the field but were fighting in the second world war.[2]

```
                    FIRST GAME
      CHICAGO (N.)              NEW YORK (N.)
            a.b. r. h. po. a  e.          a.b. r. h. po. a. e.
Hack. 3b .... 4   0 0 1  4 0   Rucker. cf ... 5   0 0 3  1 0
Hughes. ss... 3   1 0 2  2 0   Hausm'n. 3b. 4    1 1 2  6 0
Car'r'ta. 1b.. 4  1 1 12  0 0  Ott. rf....... 3   2 2 0  0 0
Nich'ls'n. rf.. 3 4 3 2  0 1   Medwick. lf.. 4   0 1 0  0 0
Dal's'dro. lf.. 4 0 1 1  0 0   Weint'b. 1b.. 4   0 1 15 0 0
Pafko. cf..... 4  0 0 0  0 0   Lombardi. c. 3    0 1 4  1 0
Johnson. 2b. 4    1 3 4  4 0   Kerr. ss ..... 4   1 0 1  5 0
Kreitner. c .. 3  0 0 5  0 0   Luby. 3b...... 3   0 0 2  3 0
Chipman. p . 2    0 0 0  0 0   Voiselle. p ... 1  0 1 0  1 0
V'and'b'g. p . 1  0 1 0  0 0   Adams.. p... 0    0 0 0  0 0
                               aSloan........ 1   0 0 0  0 0
Total         32  7 9 27 10 1  Hansen. p ... 0   0 0 0  0 0
                               bGardella.... 1    0 0 0  0 0

                               Total         33  4 7 27 17 0
aBatted for Adams in the seventh.
bBatted for Hansen in the ninth.

Chicago .............................0 1 0   1 1 1   0 1 0—7
New York ...........................0 0 0   1 1 1   0 1 0—4
```

Runs batted in—Johnson, Nicholson 4, Ott, Voiselle, Cavar-retta, Medwick, Vandenberg, Weintraub.   Two-base hit—Weintraub.   Home runs—Nicholson 3, Ott.   Stolen base—Hughes. Double plays—Hausmann, Kerr and Weintraub; Hack, Johnson and Cararretta; Rucker, Luby and Hausmann. Left on bases—New York 6, Chicago 1. Bases on Balls—Off Voiselle 3, Chipman 1, Vandenberg 2. Struck out—By Voiselle 2, Chipman 1, Hansen 1, Vandenberg 1. Hits—off Chipman 6 in 5 1-3 in-nings. Vandenberg 1 in 3 2-3 innings, Voiselle 7 in 6 (none out in seventh), Adams 1 in 1, Hansen 1 in 2. Hit by pitcher—By Vandenberg (Luby). Wild pitch—Chipman. Winning pitcher—Chipman. Losing pitcher—Voiselle. Umpires—Conlan, Barr and Sears. Time of game—2:10.

It's the first game of a double header. I found I could puzzle out the Chicago side—who scored when, driven in by whom. But when I first looked at it, I thought they had made some crazy mistakes.

Jim: The box score is organized differently, but the usual information seems to be there. I'll give it a try.

Solution on p. 221

---

[2]July 23, 1944

6.5   TWINS FALL SHORT IN SLUGFEST

Jerry: In a game on Sept. 9, 1962, Minnesota scored 9 runs scattered in 4 innings. I can figure out who scored in each inning. I can also figure out who homered when. But it's a trek!

MINNESOTA	ab	r.	h.	bi	DETROIT	ab	r.	h.	bi
Green, cf	6	0	2	0	Fernandez, ss	4	3	3	1
Power, 1b	5	1	2	0	Bruton, cf	4	2	2	1
Rollins 3b	4	0	1	1	Kaline, rf	3	1	2	3
Killebrew lf	4	0	0	0	Colavito, lf	5	0	2	2
Allison, rf	4	2	1	2	Morton 1b	2	1	1	1
Battey, c	3	0	1	0	aCash, 1b	3	0	1	1
Zim'man, c	2	1	2	0	Kostro, 3b	4	0	1	0
Allen, 2b	4	2	1	3	McAuliffe, 2b	3	1	1	1
Versalles, ss	4	1	2	0	Brown, c	5	0	1	0
Kralick, p	2	1	1	0	Bunning, p	3	1	1	0
Gomez, p	0	0	0	0	Fox, p	1	1	0	0
bMincher	1	1	1	3	Humphreys, p	0	0	0	0
Stange, p	0	0	0	0	Totals	37	10	15	10
Pleis, p	1	0	1	0					
Sullivan, p	0	0	0	0					
cOliva	1	0	0	0					
Totals	41	9	15	9					

aSingled for Morton in the 5th; bHit a homer for Gomez in the 6th; struck out for Sullivan in 9th.

```
Minnesota ............. 0 0 1  0 2 3  0 0 3—9
Detroit ............... 0 1 1  0 3 1  4 0 x—10
```
E—None. A—Minnesota 9, Detroit 14. LOB—Minnesota 10, Detroit 11. 2-B Hits—Power, Kralick, Rollins, Zimmerman, Kaline 2, Fernandez. HR—Allison, Mincher, Allen, Morton, McAuliffe. SB—Fernandez. Sacrifices—Bruton, Kostro. SF—Kaline.

Minnesota	IP.	H.	R.	ER.	BB.	SO.
*Kralick	4	7	5	5	2	0
Gomez	1	1	0	0	1	1
Stange	0.2	2	1	1	0	1
Pleis (L 2–5)	1	3	4	4	1	1
Sullivan	1.1	2	0	0	1	0
Detroit						
†Bunning	5	11	6	6	2	3
‡Fox (W, 2–6)	3	3	2	2	3	0
Humphreys	1	1	1	1	0	1

*Faced 4 batters in 5th. †Faced 3 batters in 6th. ‡Faced 2 batters in 9th. Wild pitch—Bunning. Balk—Fox. Umpires—Hurley, Flaherty, Runge, Carrigan. Time of game—2:09. Attendance—10,137.

Jim: Umm...

Jerry: This box score has something we haven't seen before: a lot of extra information in footnotes. For example, whenever there is a pinch hitter, there is a footnote that tells you who he pinch-hit for and what he did. And, when a pitcher pitches part of an inning without getting any outs, there is a footnote telling you which inning that was and how many batters he pitched to.

Jim: Why do you think they do that?

Jerry: Well, let's take the case of Minnesota's pitcher, Kralick. In the box score, he had 4 innings pitched. You might think that means he pitched for 4 complete innings and Gomez replaced him to start the fifth inning. The footnote tells you that's wrong—actually he pitched to 4 batters in the fifth. Since he got none of them out, he still only has 4 innings pitched.

Jim: That makes sense. You wouldn't want to mislead the reader.

Jerry: But actually, we don't need the footnote to conclude that Kralick pitched in the fifth inning!

Jim: Why not? ... Oh, I see. Kralick gave up 5 runs, but Detroit got only 2 runs in the first 4 innings. Kralick must have pitched to at least 3 batters in the fourth inning to be responsible for 3 more runs.

Jerry: Actually, I find these footnotes a bit annoying (the ones in box scores, not the ones in this book). Often you don't need them at all to figure out the important things that happened.

    For an extra challenge, see if you can figure out for Minnesota: who scored in each inning and who homered when (with proof), without using the footnotes!

Hint on p. 172; solution on p. 223

> Sometimes box scores include footnotes for each pitcher who pitches part on an inning without getting any outs. The footnote says which inning this is and how many batters he faced. But these box scores never have footnotes that tell you how many batters were faced in an inning by a pitcher who did get somebody out in the inning.

6.6 IN SCORING POSITION

TWINS 11, INDIANS 10

Cleveland	AB	R	H	BI	BB	SO	Avg
Lofton cf	4	1	2	0	0	0	.266
b Cabrera ph ss	1	1	1	0	0	0	.333
Vizquel ss	4	1	1	0	0	1	.254
c WCordero ph	0	1	0	0	1	0	.318
RAlomar 2b	4	1	1	4	0	0	.343
JGonzales rf	4	2	3	2	1	0	.359
Thome 1b	4	1	3	3	1	1	.285
Burks dh	4	1	1	1	1	1	.309
MCordova lf cf	5	0	0	0	0	0	.353
Fryman 3b	5	1	2	0	0	2	.222
Taubensee c	3	0	0	0	0	1	.242
a EDiaz ph c	2	0	0	0	0	1	.313
Totals	40	10	14	10	4	7	

Minnesota	AB	R	H	BI	BB	SO	Avg
J.Jones lf	6	1	2	0	0	3	.264
CGuzman ss	6	3	4	2	0	1	.296
Mientkiewicz	4	1	1	0	0	0	.344
Lawton rf	5	2	2	3	0	0	.290
THunter cf	5	0	1	1	0	2	.261
Koskie 3b	2	1	1	0	3	1	.245
Buchanan dh	4	1	1	2	1	2	.212
Pierzynski c	4	0	2	2	0	2	.274
d Hocking ph	1	0	0	0	0	0	.269
Rivas 2b	5	2	2	0	0	1	.221
Totals	42	11	16	10	5	13	

E—Vizquel (4), Taubensee (2), Rincon (1), Pierzynski (5). LOB—Cleveland 8, Minnesota 11. 2B—Thome (10), J.Jones (9), Mientkiewicz (15), Lawton 2 (12) Buchanan (2). 3B—Vizquel (2). HR—Burks (13) off TMiller, Thome (14) off Mays, JGonzalez 2 (15) off BWells, Mays, RAlomar (5) off Mays. RBIs—RAlomar 4 (34), JGonzalez 2 (52), Thome 3 (35), Burks (40), CGuzman 2 (16), Lawton 3 (23) THunter (30), Buchanan 2 (11), Pierzynski 2 (15). SB—Lofton (7), Koskie (7), Rivas (9). SF—RAlomar. Runners left in scoring position—Cleveland 4 (Burks, Fryman 3) Minnesota 7 (JJones, Lawton, Buchanan 2, Pierzynski 3).

Cleveland	IP	H	R	ER	BB	SO	NP	ERA
Burba	1.2	7	8	6	3	3	54	6.02
Westbrook	4.1	3	0	0	0	2	76	3.86
Rincon	1	3	2	1	0	1	20	4.05
Shuey L	1.1	3	1	1	0	3	26	2.43

Minnesota	IP	H	R	ER	BB	SO	NP	ERA
Mays	.6	9	5	5	2	4	108	2.98
BWells	0.1	1	0	0	0	0	5	4.06
TMiller	0.1	1	1	1	0	1	11	5.00
Cressend	0.1	1	1	1	0	0	12	3.09
Guardado W	2	2	2	2	2	2	30	4.85

Mays pitched to 1 batter in the 7th. Cressend pitched to 1 batter in the 8th. Inherited runner scored—Westbrook 2 2, BWells 1 1, Guardado 1 1. IBB—off Guardado (JGonzales), 1 off Westbrook (Koskie) 1. Umpires—Wieters, Barrell, Ted, Marquez, Rippley. T—3:36. A—20,617.

Indians . 2 0 1 0 1 0 3 3 0—10 14 3
Twins .. 2 6 0 0 0 0 2 0 1—11 16 1

One out when winning run scored. a struck out for Taubensee in the 8th. b singled for Lofton in the 8th. c walked for Vizquel in the 8th. d grounded out for Pierzynski in the 9th

Jim: This is cool! For this game, which is more recent than the previous games,[3] the box score discusses players left on base "in scoring position" (second or third base) at the end of an inning.

---

[3] June 4, 2001

Jerry: "Runners left in scoring position—Cleveland 4 (Burks, Fryman 3)" I know that 'scoring position' means second or third base. Does this mean that Burks and Fryman got left on base in scoring position?

Jim: I looked it up. You leave someone in scoring position if you make the last out in an inning and there is a teammate on second or third. Burks and Fryman weren't left on base. They left others on base.

There's an acronym, RLSP.

Jerry: Will that help?

With all the data in the box score it's possible to:

1. Determine when Burks and Fryman left runners in scoring position.
2. Determine who drove in each run.
3. Determine when and by whom each run was scored in the seventh and eighth innings.

It's tricky.

There are hints on page 173 and the solution is on page 225.

CHAPTER 7

More Puzzling Art

7.1 JUST TOTALS

Jim: Jerry, you and I have different puzzle aesthetics. I'm attached to clean, spare presentations, with hidden depths. You like sprawling, giant puzzles, puzzles that seem to encompass the universe of baseball.

Of course, we both appreciate puzzles of all sorts.

Thinking of puzzles as art, I would describe my sort of puzzle as "Classical." And the word for the sort you especially like might be "Baroque." What do you think?

Jerry: I was thinking "Minimalist" and "Heavy metal."

Jim: Oh.

Jerry: But really, we each appreciate both directions in constructing puzzles. Here, for example, is a minimalist puzzle I just put together:

The White Sox played the Indians in Cleveland (real game). Here are the non-zero totals for the two teams together:

AB	R	H	BI	BB	SO	DP	2B	3B	HR	LOB	SB
93	5	19	5	8	31	4	4	1	2	18	2

DOI: 10.1201/9781003332602-7   **97**

Jerry: Who won?

Jim: Beautiful!

Solution and a bonus puzzle on p. 229

7.2 TWO GUYS

Jerry: Back to the Cats and Dogs. Here are some stats for the Cats'
4–2 win.

Cats

	ab	r	h	bi
Abe ss	4	0	1	0
Brown 2b	4	0	1	0
Castro rf	4	0	1	0
Dominguez lf	3	0	2	4
Engel c	5	0	0	0
Frank cf	4	0	1	0
Guzman 3b	4	0	0	0
Hall 1b	2	4	2	0
Iglesias p	3	0	1	0
Totals	33	4	10	4

Jim: And you want to know who scored when, etc?

Jerry: No no. Just tell me who led off for the Cats in the sixth
inning. And tell me how many they left on base in the
seventh.

Jim: Those are weird questions to ask. And why is this chapter
called "Two Guys"? Who are the two guys?

Jerry: Good questions!

Hint on p. 173; solution on p. 230

7.3 SMEDLEY

Jim: This is another puzzle about failure. Consider this. Smedley is the starting shortstop for his team. In one 9-inning game, Smedley was up three times. Every time he batted, he had the last plate appearance of the inning and he struck out. His final plate appearance was the last of the game. Smedley was bad news.

Smedley's team left just 2 men on base. What can you say about this game?

Jerry: Okay, this going to be tricky. You're not asking who won, right?

Jim: How could I ask that? I've given you no information about the other team!

Jerry: True. No information about the other team. Still ...

Solution and a second bonus puzzle on p. 231

7.4 EVERYBODY SCORES A RUN

Dogs	ab	r	h	bi
Amsler	6	1	1	0
Blanco	7	1	6	6
Clark	6	1	5	0
Durand	3	1	2	1
Emerson	5	1	2	1
Foster	5	1	2	1
Gato	5	1	1	0
Herrera	6	1	1	0
Ito	5	1	1	0
Totals	48	9	21	9

Dogs   1 1 1   1 1 1   1 1 1--9

LOB--Dogs 27, HR--Blanco,
3B--Amsler, 2B--Clark, Durand,
Emerson, SF--Blanco, S--Herrera,
E--Cats, maybe

Jim: I'm supposed to sort this all out?

Jerry: Of course!

Solution on p. 232

7.5 CASTRO HOMERS IN THE FIRST

Jerry: Here's another one.

It's all about a team effort. Not about you and me. It's the guys on the team. Everybody contributed.

The fun started in the first inning when Castro, batting third for the Cats, slugged a home run. The Cats scored in each of the nine innings. Every player and every inning had 1 run, 1 hit, and 1 rbi. Six of the hits were homers. The Cats left nobody stranded on base. The last batter for the Cats got a hit.

Jim: A real team effort! What's the question?

Jerry: Who won? And who scored in each inning and who drove him in? That should keep you busy!

Jim: We have zero left on base. We have 1 run each inning.

So each inning had 4 batters; 36 plate appearances in 9 innings.

So the #9 batter, the pitcher, I guess, batted last in the game. And it was a hit.

Jerry: Maybe. Maybe not. But if the last batter got a hit in his last at-bat, how did the game end?

Jim: The guy must have made an out after his hit. He could have been picked off by the pitcher. Or he might have tried to stretch a single into a double. His out was the third out, and was made after the run in the ninth inning was scored.

That tells us who won the game: it can't be the Cats, because then when the run was scored, the Cats would immediately have won in a walk-off and there would never be a third out. So the Dogs must have won the game. But I have no idea who the last batter drove in. It could have been any of the three preceding batters.

Jerry: How do you know for sure that the ninth player batted at all in the ninth inning? You just said that if the Cats won, it would be a walk-off win, with fewer than 3 outs. So couldn't the Cats have had only 1, 2, or 3 batters up in the ninth, and the pitcher never reached the plate?

Hint on p. 173; solution on p. 235

7.6   BROWN HOMERS IN THE FIRST

Jim: I liked that last puzzle. I liked it so much I constructed another just like it. It's the exact same problem except that it is Brown, batting second, who homers in the first inning, rather than Castro.

Solution on p. 237

7.7 LEFTOVERS

Jim: Jerry, I got a puzzle in the mail! It's a lot of pieces of paper!

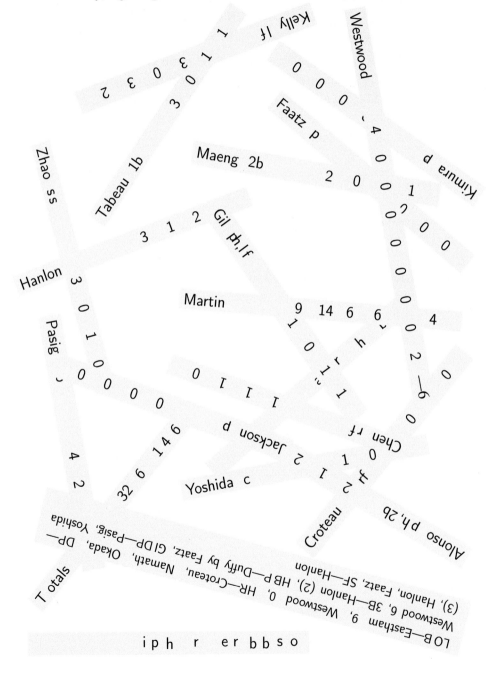

Jerry: Who sent it?

Jim: There's no return address.

Jerry: Suspicious.

Jim: Maybe it's from Nancy!

Jerry: Ridiculous.

Jim: I think it's the other half of the game that you invented in Chapter Five. Did *you* send this?

Jerry: **No.**

Jim: I think it's another masterpiece!

⟨Long, lingering, pause⟩

Jerry: We'll see.

Hint on p. 174; solution on p. 238

CHAPTER 8

Missing Information

8.1 FOUR PLAYERS MISSING

Jim: Some of the coolest puzzles are ones where you're given very little information but you discover deep facts. A great puzzle is like a poem, a delicate work of art, a jewel that opens up worlds. It's hard to sculpt such problems.

Jerry: But worth the effort!

        What have you got?

Jim: Just a little thing:

	ab	r	bi
1.	4	0	0
3.	3	1	0
5.	4	0	0
7.	3	0	1
9.	3	0	0
	30	2	2

```
1 1 0  0 0 0  0 0 0--2
```

        Who scored, when they did they score, and who drove them in?

Jerry: What?!?

        Okay. Since you put 9 at the bottom, I'm assuming there are only 9 batters, so there are no substitutions.

        So I guess we are missing the stats for the batters in slots 2, 4, 6, and 8. Our job is to figure them out. Right?

DOI: 10.1201/9781003332602-8

Jim: Right.

Jerry: This isn't real, of course. But since it's fiction, shouldn't
         there be a story? Why is the box score missing stuff?

Jim: Oh yeah. My story is that the sports editor of the city's paper
         is determined to print box scores for all the high school
         games in the greater metropolitan area. But the editor is only
         given half a page for all high school sports news. So she's
         been using abbreviations.

         I enlarged it for you.

Jerry: Uh, thank—

Jim: This is how it actually appeared in the paper:

	ab	r	bi
1.	4	0	0
3.	3	1	0
5.	4	0	0
7.	3	0	1
9.	3	0	0
	30	2	2

1 1 0   0 0 0   0 0 0--2

Solution on p. 242

8.2   FOUR MISSING, AGAIN

Jim: If you liked that puzzle, here's another.

	ab	r	bi
1.	4	0	1
3.	3	0	2
5.	4	0	0
7.	3	0	1
9.	3	0	1
	30	5	5

1 0 0   3 1 0   0 0 0--5

Jim: Also no substitutions in the game.

         Who scored, when they did they score, and who drove them
         in?

Solution on p. 242

8.3   SIX MISSING

Jerry: That was pretty easy. But I really like the idea of missing information so I wrote a puzzle too. You don't want many clues, right? So how about this?

	ab	r	bi
?. Alice		4	
?. Bryanna		4	
?. Chrissy		4	
		12	

2 2 2 2 2 2 0 0 0--12

Jerry: In each inning, who scored the runs? And in which batting slots did Alice, Bryanna, and Chrissy bat?

Jim: You're not giving me any information about where the players hit in the lineup?

Jerry: Well, they're in that order, that is, Alice bats before Bryanna (but not necessarily right before Bryanna), and similarly Bryanna bats before Chrissy. But you can assume that there were 9 batters in total, with no substitutions.

Jim: Gosh. Hardly any clues.

Jerry: Actually, I gave you more information than you need.

Jim: That's exciting!

        Can I call these guys[1] A, B, and C?

Jerry: If that makes it easier for you.

Jim: Um ... yes.

        Well, between any two scorers in an inning, there can't be more than two non-scorers. It's the Gap rule (p.24). So the gap from A to B is 0, 1, or 2. And the gap from B to C is also 0, 1, or 2.

[1] "Guy," in contemporary usage, is not gender-specific.

Hey! The gap from C to A is also 0, 1, or 2! And the only way that can happen is if they're all equally spaced, with gaps of 2!

And then also B must be in one of the first four slots or else there will be 3 outs before B comes to the plate. So here's your answer:

AxxBxxCxx

That wasn't so hard!

Six up the first inning. That's A and B scoring. Six up in the second inning with C leading off. That's C and A scoring. Then six up in the third inning with B leading off. That's B and C scoring and we're ready to repeat because A will lead off the fourth!

Jerry: You've found one solution. Could there be others? You've made a lot of assumptions. Here's one: you assumed A and B score together twice, B and C twice, and C and A twice.

Jim: Ooh.

Well, say A and B score x times, B and C score y times, and A and C score z times. That gives us: $4 = x + z$ (A scores 4 runs) and $4 = x + y$ (B scores 4 runs). So $x + z = x + y$. So $z = y$. Similarly we can get $x = y$. They're all equal. So they all have to be 2.

Jerry: Good.

Jerry: Now I agree that two of these guys have to be in the first 4 slots, so you gotta have A and B there and the gap from A to B is 0, 1, or 2. But what about C? You assume the gap from C to A is 0, 1, or 2. But couldn't it be that the gap is from A to C, not from *C to A?*

Jim: That's crazy! The only way that could happen is if A, B, and C were all bunched together in four slots! Like AxBC!

Jerry: Why couldn't you have a bunch in 5 or 6 slots? Like AxBxC? Or AxxBxC?

Jim: Then A and C would be too far apart for both to score in the same inning no matter which one batted first! The gap is more than 2 slots in either direction.

Jerry: Okay, I see it would have to be in 4 slots. So prove to me that's impossible.

⟨Long pause here. More than 24 hours, actually.⟩

Jim: Okay! Here are the 4 slots (in gray):

Since the first player to score in any inning must be one of the first 3 batters, each inning will have to start by leading off in one of these five places:

| | | ▌|▌|▌|▌| | |
∧ ∧ ∧ ∧ ∧

Jerry: Okay.

Jim: And you can't keep this up for 6 innings.

Jerry: Why not?

Jim: Say someone leads off in one of those spots. The inning has 2 runs so it will have 5–8 plate appearances. 5 is the minimum, so the next person leading off is beyond the 5 slots. The only hope is that the inning wraps around to one of those slots.

Jerry: Okay.

Jim: But even with 8 plate appearances, the person leading off next inning will be 1 slot to the left. That might work except that you can only do that four times. So you can't lead off from one of the 5 slots in six consecutive innings.

        Ta Da!

Jerry: Brilliant!

Jim: Now we have it. The gap from A to B and the gap from C to A is 0, 1, or 2. And the gap between B and C? It can't be from C to B, because that would bunch them up again, like CAxB. And I already showed that's not possible. So the gap has to be from B to C. And we're done.

8.4 FIVE MISSING

Jerry: Fun, right? Here's another puzzle:

	ab	r	bi
?. Alice		2	
?. Bryanna		2	
?. Chrissy		3	
?. Debbie		0	3
		7	

```
2 2 2  1 0 0  0 0 0--7
```

Jim: These puzzles are so ... bizarre!

         This looks a lot like the Six Missing problem.

Jerry: Aha.

Jim: In that problem, making A, B, and C equally spaced apart worked:

                      AxxBxxC.

Jim: Let's try it here. A has to lead off because two ladies must score in the first.

                      AxxBxxCxx.

So A and B score in the first. Six up. And then C and A score in the second, six up. Then B and C score in the third ... how many up?

But in this problem, C has to score in the fourth. In the third, B leads off. At most 8 come to the plate (3 LOB). That means slot 3 leads off the fourth. C is fifth up in the inning. She can't score!

	1	2	3	4	5
A	╳		•		
	•	•	•		
	•	•		?	
B	◇		◇	?	
	•		•	?	
	•		•	?	
C	╳		╳	?	
		•	•		
		•	•		

Jerry: Yup. Doesn't work.

Jim: Well, alright then, I'm going to guess that C, who scores three times, is the gal who scores in the fourth. How do I prove that?

Jerry: Show it's impossible for C to score in each of the first 3 innings.

Jim: Well of course! So ... to score in the first, she has to be among the first 4 slots. But she has to follow A and B, so that means slot 3 or 4. There can be at most 8 PA in the first, so the second inning starts with the 9 slot at the latest. Now for C to score, she has to be in slot 3.

Ah! And then the third inning starts with the 8 slot at the latest. So C comes up fifth in the inning and can't score!

Jerry: Good. You win. C can't score in each of the first 3 innings. She has to score in the fourth. That means that the first 3 innings are just like the Six Missing problem, A and B in 1 inning, A and C in another, and B and C in another. We figured out then that this means that if they aren't spaced out AxxBxxCxx, then A, B, and C have to be in a 4-slot block.[2] So that's true here.

Jim: That means either ABCx, AxBC, or ABxC.

⟨pause⟩

Jim: This puzzles me. C fails to score in one of the first 3 innings. Does that mean that in one of her early at-bats she doesn't score? So that in the fourth, she scores in her fourth at-bat?

Jerry: Do the numbers.

---

[2]p. 108

Jim: Right. Okay, in the first 3 innings, we have at most 24 PA (8 each inning). In the fourth, C must come up in one of the first 4 at-bats, so at the latest, the 28th plate appearance. But that can't be her fourth at-bat unless she's in slot 1, which she isn't.

So she has to score her third run in her third plate appearance. She can't ever miss a chance to score.

Now what?

Jerry: So check out ABC.

Jim: Yes, ABCx. Hey, that's easy. In one of the first 3 innings, A and B are the ones who score. B doesn't end the inning. So C has to bat and must miss a chance to score. No good!

And that same reasoning shows that AxBC won't work either.

We're done! ABxC is the answer! Put D after that, ABxCD!

Jerry: You haven't shown that ABxC works. Only that those others fail. Show me ABxC works.

Jim: Piece of cake. Here goes:

In the first, one of A, B scores. Then C scores.

She can't score in the second—at best, she's fifth up in the inning. So the inning has to end before she comes up ... A and B score ... then slot 3 makes the third out ...

	1	2	3	4	5
A	◇	◇			
B	◇	◇			
	•	•			
C	◇				
	•				
	•				
	•				
		•			
		•			

Ah! This works if just 7 players come up in the first.

Okay, there's a problem. C can score in the third, but no one else can! Bad news!

Did you know this was going to happen? We're sunk! This puzzle has no solution!

Jerry: There is a solution. You've overlooked something.

Jim: No I haven't! I'm a really careful thinker. Nothing escapes me.

⟨pause⟩

Jerry: You know, sometimes a person sees something and that enables the person to turn the corner and go on. This puzzle puts you in a different position. You have to turn the corner *first* and *then* you see something.

⟨long pause⟩

Jerry: To put it another way, ABxC isn't the only way C can be in a block of four with A and B.

⟨long pause⟩

Jim: OMG. The "corner" is the end of the batting order. If C is all the way at the end, then turning the corner, she's next to A! It's

ABxxxxxxC !

No wait. D has to come after C. It's

ABxxxxxCD !

Jim: It's a block of four. It's basically CDAB! Okay. I've got it now. A and B score in the first. C and A score in the second. Then C and B ... wait. That's not going to work. B's fifth up in the inning.

Now I've really got it! In the second it's C and B scoring (instead of C and A) and then C and A score in the third. Then finally C scores in the fourth.

And all along the way, D can drive C in.

8.5 WHO WON?

Cats	ab	r	h	bi
Abe				
Brown				
Castro				
Dominguez				
Engel				
Frank				
Guzman				
Hall				
Iglesias p	1	0	0	0
Ivory ph	1	0	1	1
Ingals p	1	0	0	0

	ip	h	r	er	bb	so
Cats						
Iglesias	5	7	3	3	1	2
Ingals	4	3	3	3	1	4
Dogs						
Ito	5	2			0	15
Ignacio		4			3	2

LOB—Cats 2, Dogs 8, DP—Cats 2, 3b—Hall, 2b—Emerson

Jim: Dogs and Cats. This is pretty simple. All I want to know is who won the game?

Jerry: But you're giving me practically nothing!

Jim: I've given you everything except for the missing numbers in the batters' and pitchers' lines.

⟨There is a pause here.⟩

Jerry: Hey! I got it! The Cats are the visitors, since their pitchers are listed before the Dogs'. Since the Cats pitchers have just 9 innings pitched, the Dogs batted in only 9 innings. But the Dogs had a complete ninth inning, getting all 3 outs. So they must have been behind or tied at the end of the 9th. They can't have been tied, because then they would have had to bat again in the tenth. So they lost!

Jim: Nope. You forgot that a team credited with pitching just 9 innings might still have pitched in a tenth—as long as they got no outs in the tenth. Having an ip of 9 just means getting a total of 27 outs. It's not the number of innings the guy pitched in. The Dogs could have scored in the bottom of the tenth with no outs, getting a walk-off win.

Jerry: Of course! .... I mean, I knew that.

But now what?

Hint on p. 175; solution on p. 243

8.6 SAD STORY

Jerry: This is all about failure. It looks like success, but it's failure.

Dogs Batting

	ab	r	h	bi
Amsler		1		
Blanco				
Clark				
Durand		1		
Emerson	4	2	4	1
Foster				1
Gato				2
Herrera				
Ito				
Totals				

```
Cats ··  0 0 0  1 2 0  0 1 1--5
Dogs ··  1 0 1  0 0 0  0 2 0--4
```

Cats Pitching

	ip	h	r	er	bb	so
Iglesias (W)	8		4	4	0	6
Ivory	1	0	0	0	1	2

LOB- Dogs 10, HR-Emerson, 3b-Emerson,
2b-Emerson, Gato, Amsler, SB-Emerson (2)

Jerry: In this game, Emerson hit for the cycle (single, double, triple, home run). That's a big deal.[3] In addition to that, he stole two bases. Despite this, he considered himself a failure.

My puzzle is to find out exactly when Emerson performed each of his feats.

Jim: Okay, this going to be tricky. You're not asking who won, right?

---

[3]Hitting for the cycle is about as rare as pitching a no-hitter—in major league history—as of this writing—there have been 333 cycles and 313 no-hitters. But chance plays a big role here. Thirty-three players have had more than one no-hitter; only one man has hit for the cycle twice.

Jerry: Yeah, because it's right there in the box score!

Jim: And you're not asking why he thought he was a failure.

Jerry: How could I ask that? That's psychology. We only deal here with strictly logical questions!

Jim: True. The soul of man is a far country.[4] Still ...

Solution on p. 245

---

[4]D. M. Thomas

CHAPTER 9

Little League Games

9.1 KIDS' STUFF

Jerry: Hey, Jim, I have an idea for another kind of puzzle. Did you ever play in Little League?

Jim: Yeah, but I wasn't much good. I always struck out.

Jerry: Well, don't feel bad. I failed the tryouts and never got to play at all! But I've been to Little League games, and it seems that practically every batter either strikes out or walks. So my idea for a puzzle is to have a game in which *everyone* either strikes out or walks. Here's an example:

It's a 6 inning game, with a single run each inning,

$$1\ 1\ 1\ 1\ 1\ 1 - 6,$$

and every batter is either out on strikes or reaches first on a walk.

	ab	r	h	bi
Aiko	2	1	0	0
Butch	2	0	0	2
Chulo	2	1	0	1
Dusty	2	1	0	0
Em	2	0	0	1
Fei Fei	2	1	0	1
Gordito	2	0	0	1
Hank	2	1	0	0
Izzie	2	1	0	0
Totals	18	6	0	6

DOI: 10.1201/9781003332602-9

Jerry: The challenge is this: recreate the entire scorecard. In each inning, for each batter: figure out who walked, who struck out, who scored, and who drove in whom.

> To make sure you get started out on the right foot, we're offering two hints this time, one on p.175 and one on p.177. But maybe you won't need either of them.
> The solution is on p.246

9.2 KIDS WILL BE KIDS

Jerry: I've got another Little League puzzle. It's harder, I think. This time the line score is 1 6 4 1 2 1 — 15 runs for the losing team (the Puppies). The winning team (the Kittens) got even more runs.

Jim: I guess in these games the umpires have to have a very big strikeout zone or the game will never end. The batters have to swing at the bad pitches. No wonder they never hit the ball!

Jerry: Here are the batting lines for the Puppies:

	ab	r	h	bi
Aiko	1	4	0	1
Butch	2	2	0	1
Chulo	3	3	0	0
Dusty	2	0	0	0
Em	1	1	0	4
Fei Fei	4	2	0	1
Gordito	1	1	0	4
Hank	2	1	0	2
Izzie	2	1	0	2
Totals	18	15	0	15

Jerry: Since we're given that every batter either walked or struck out, the at-bats must all be strikeouts. There weren't any other outs. So no batter who walked got thrown out on the bases.

And there were 15 rbis. You can't get an rbi on a strikeout, so all the runs were scored on bases-loaded walks, not from wild throws trying to catch somebody stealing a base or for any other reason.

Again, recreate the entire scorecard. In each inning, for each batter, who walked, and who struck out?

Hint on p. 176; solution on p. 248

9.3   MACY AT THE BAT

Jim: I'm going to need your help on this.

Jerry: A puzzle isn't working out?

Jim: No the puzzle works. But I need a story. I mean, the puzzle looks just a little ridiculous.

Jerry: Ridiculous? That never stopped you before.

Jim: Here it is: In this game no player on either team ever had different at-bats. Say, for example, Macy was the third to bat for his team in the first inning. When Macy first comes to the plate there is one out and Stewie is on second. Macy homers. Then every other time Macy comes to the plate, Macy is the third to bat in the inning, there is one out, Stewie is on second, and Macy homers again.

It's like that in this game for every player on both teams. Everybody's at-bats keep repeating. Ridiculous, right?

Also, this is a 6-inning game—not shortened by weather or anything. Macy's team scored 2 runs.

The question: Did they win?

Jerry: This looks like a lovely problem. I think I'm going to like it. But it does sound pretty weird.

Solution on p. 250

9.4 IT'S AN ERG

Jerry: I've got it. I know how to explain that last puzzle of yours.

Jim: Where every at-bat is repeated?

Jerry: Yeah. When you said "Little League" I began thinking of Taiwan, and international games. You know they train Little League teams very hard over there. So imagine that there's this coach in Taipei who has a special training technique. The coach starts a training game normally, but then if anything happens differently, they do the at-bat over and over until it successfully repeats. The kids learn bat control, learn how to hit to certain fields, the fielders learn to anticipate, and so on. They call these games "ERGs," for "Eerie Repetition Games." Do you like that?

Jim: Ooh.

Yes! We could say this is one of the reasons that teams in Taiwan have been so successful internationally in Little League contests.

I like it so much I think it's real![1]

Jerry: And I dreamed up an ERG. Like yours, it's a 6-inning game. Taipei scores 8 runs.

Jim: That's all?

Jerry: That's all. Who won?

Solution on p. 251

---

[1] It isn't real.

9.5 ZHANG AND LI

Jim: Jerry, what if an ERG ends with a walk-off win? Say a player makes a walk-off single. The last time this happened, the next batter came to the plate and batted but this time the game is over. Does that violate our rules for what an ERG is?

Jerry: No no. The rule is (you can look it up) that—

Jim: I can't look it up! We haven't written anything down!

Jerry: —the rule is that players must eerily repeat until the game ends. Whenever that is.

Jim: Okay. Write that down.

Jerry: Later—I have a new puzzle! The coach who invented the training method is the manager of the Taipei Tigers. Taipei had a training session with Taoyuan yesterday.

Jim: Oh boy!

Jerry: They played a doubleheader, two regular games (6 innings, not rain-shortened). All of the runs for the Tigers that day were driven in by Li, who bats second in the lineup, and all of the runs for the Dragons were driven in by Zhang, who bats seventh. The two games had identical scores, except that the Tigers won one, and the Dragons won the other. In each game, the winning team won by a single run. It took 10 hours for the kids to make it happen.

Jim: And the question?

Jerry: What were the scores? And whose park did they play in?

Hint on p. 178. Solution and a bonus puzzle on p. 251

9.6 LI AND ZHANG

Jim: Jerry, isn't it possible that some of these games are infinite? What if each team scores one run every inning?

Jerry: That never happens, because we just made this up.

Jim: We made it up. So it's ours. That means that if something goes wrong, we're responsible!

Jerry: Infinite games are theoretically possible in regular baseball, right? Who's responsible for that?

Jim: Oh wait.

I got it. ERGs are superior to reality! Both real and ERG games can be infinite in theory, but with ERGS we can know ahead of time that they're going to be infinite!

Jerry: Aha! Okay, in an ERG, as soon as the coaches know the game will be infinite, they quit and start a new game. Problem solved.

Jim: Excellent! And I have some news. Li and Zhang just played another double header. (Those kids must be exhausted.) Again, they were regular games, not rain-shortened, no extra innings. And again the teams split. And again Li and Zhang drove in all the runs (Li batting second and Zhang batting seventh) 8 runs total for each of them.

Jerry: And the question?

Jim: As before, what were the scores?

Solution on p. 253

9.7 EXTREME ERGS

Jerry: Jim, What is the longest possible completed ERG game (in terms of plate appearances)? And what is the longest possible extra-inning ERG?

⟨An hour goes by.⟩

Jim: Got them. And here's a question for you: What's the shortest possible ERG?

Solution on p. 254

9.8   ADVANCED ERGONOMICS

Jerry: Here's another Little League 6-inning ERG game: The Taipei Tigers had 8 more plate appearances than the Hong Kong Dragons. The game wasn't shortened by rain or for similar reasons. What was the final score?

This will take some deep ERGonomics.

Just a warning.

What was your major in college?

Jim: Thanks.

Hint on p. 178; solution on p. 255

The Final Problem

10.1 CHILD'S PLAY

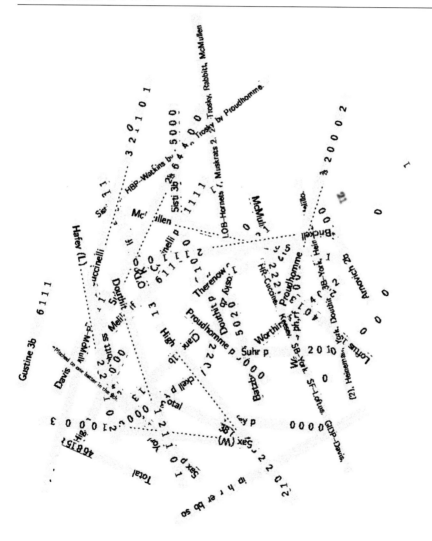

DOI: 10.1201/9781003332602-10

Jim: Wow! Jerry, you've done it again! Another masterpiece! All those players! There's at least two teams here! Wow wow wow!

Jerry: **NO.**

I mean, "No."

I mean, I haven't done anything.

This is just ... a minor annoyance from my sister.

Jim: Nancy sent this?

Jerry: Yeah.

Her kids were cutting up sheets of construction paper. Nancy came across an old newspaper that was used to pack something. It was a sports section so she had the kids cut up a box score. She told them:

"Uncle Jerry can put any box score back together!"

Jim: But we can't! It's only special box scores that we can ... that we can ... that we can make deductions about.
Does she really think we can put all those scraps together?

Jerry: No. She knows it's ridiculous. She teasing me. She's always teased me.

Jim: Oh.

We don't know anything. We don't who is on which team.

Do you know any of these players? This game could have been played years ago.

Jerry: No, I don't know who they are.

Jim: Have you tried to make any sense of this?

Jerry: Well, not yet. I was going to make a quick stab at it and then mail it back.

⟨pause⟩

Jim: Well for one thing, I see only two catchers. They must be on opposite teams.

Jerry: Oh that's right.

Jim: What are all those numbers?

Jerry: Yeah, I was wondering about that.

Jim: You don't suppose those are the cut-up line scores?

Jerry: Oh. Maybe. What a mess.

⟨pause⟩

Jim: Well, suppose we try to organize the pieces?

⟨20 minutes pass⟩

Jerry: Okay. Here are the pitchers. In alphabetical order.

	ip	h	r	er	bb	so
*Brickell	2	1	1	1	0	0
Cuccinelli	3	2	1	1	0	1
Douthit	1	2	1	1	0	1
Hafey (L)	2	2	1	1	0	2
High	4	5	2	2	0	0
McMullen	4	6	5	5	3	1
Proudhomme	⅔	2	0	0	0	2
Sax (W)	3	1	0	0	0	3
Suhr	2⅓	6	4	4	0	0

And here are the position players. In their positions.

	ab	r.	h.	bi
Clancy 1b	5	1	3	2
Loftus 1b	4	0	1	3
Trosky 2b	4	2	2	0
Arnovich 2b	3	1	0	0
Gustine 3b	6	1	1	1
Sisti 3b	5	0	0	0
Rabbitt ss	4	1	2	0
Batzberger ss	5	0	2	0
Heineman ph,ss	2	0	2	2

Melillo lf	6	1	1	0
Keesey lf	5	1	1	1
Gruhe cf	6	0	0	0
York cf	3	4	2	0
Watkins ph,rf	0	0	0	1
Worthington rf	4	0	2	2
Davis rf	5	0	1	1
O'Doul c	5	0	0	0
Therenow c	4	0	2	1

And the pitchers bat too:

Brickell p	0	0	0	0
Cuccinelli p	1	1	1	1
Douthit p	1	0	1	0
Hafey p	0	0	0	0
High p	1	0	0	0
Proudhomme p	0	0	0	0
McMullen p	2	2	2	0
Sax p	1	0	0	0
Suhr p	2	0	1	0

Then there are the lines of extra stuff. It was pretty easy to put the lines together. Doubles are always before triples and triples before homers ...

LOB—Hornets 7, Muskrats 2. 2B—Trosky, Rabbitt, McMullen (2), Heineman, York, Douthit. 3B—York, Heineman, Melillo. HR—Cuccinelli, Keesev. SB—York. SF—Loftus. GIDP—Davis, Sisti (2). HBP—Watkins by Hafey, Trosky by Proudhomme. B—McMullen.

Jim: Hornets? Muskrats?

Jerry: I guess it's a minor league game.

And finally, there are all these snippets of numbers:

21 13 3 2 1 1 1 00 00
0 0 0 0 0 0 0 0 0

Jim: Wait. There's an asterisk by Brickell. What's that?

Jerry: Oh. Hold it. There was this extra thing. It's got an asterisk too. It must go with Brickell:

*Pitched to one batter in the 9th.

I hope we haven't lost anything.

Jim: Hmm...

Sax, the winning pitcher, and Hafey, the losing pitcher, are certainly on different teams.

Suppose we total the batting and pitching numbers for the two teams together. Maybe that will tell us something.

	ab	r	h	bi
Totals:	84	15	27	15

	ip	h	r	er	bb	so
Totals:	22	27	15	15	3	10

Fifteen runs. All of them driven in. All of them earned runs. And 22 innings. That means each team had 11 innings—11 full innings—no walk-off. So the visiting team won.

That's something. Maybe we can do a bit more.

⟨long pause⟩

Jim: Hey, c'mon Jerry! Those kids look up to you.

Jerry: *Stop that.*

Jim: We did a little.

Jerry: Not much. She's teasing us.

Well, she's teasing *me.*

Jim and Jerry made more progress. They sorted out the pitching. See what you can do. Or move to the next page. It's your choice.

10.2   PITCHING IN

Jim: Fifteen runs. How were they split between the teams? We'd know if we knew how the pitchers are split up.

Jerry: Remember that footnote that says Brickell pitched to 1 batter in the ninth? These footnotes are used only when a pitcher gets no outs in the inning. So Brickell couldn't have finished the ninth inning—some other pitcher must have replaced him in the ninth. That means his 2 innings pitched must have been the seventh and eighth innings.

Jerry: Now there have to be pitchers on Brickell's team who can pitch the first 6 innings. They can't be Hafey or Sax, the losing and winning pitchers, who must have pitched in the eleventh. The only set of remaining pitchers whose ips add up to six is Suhr, Proudhomme, and Cuccinelli.

Pitchers	ip	r
Brickell	2	1
Cucc.	3	1
Douthit	1	1
Hafey (L)	2	1
High	4	2
McMul.	4	5
Proud.	.2	0
Sax (W)	3	0
Suhr	2.1	4

Five runs are charged to those pitchers, plus one more run to Brickell. Six runs in all.

Jim: That takes care of the pitchers on Brickell's team through the first 8 innings. The last 3 innings could be pitched by Sax (3 ip) or Hafey and Douthit (2 ip and 1 ip). Was it Sax?

Jerry: Sax gave up no runs. That would mean Brickell's team gave up only 6 runs. They would win 9–6.

Jim: So they scored 3 runs in the eleventh inning! How did they . . .

Jerry: No, that's not possible. The losing pitcher, Hafey, pitched the last 2 innings and gave up just 1 run. It's got to be the other way around. Hafey and Douthit go with Brickell. Brickell's team gives up a total of 8 runs. They lose.

Jim: Then we have it! The losing team's pitchers are Suhr, Proudhomme, Cuccinelli, Brickell, Douthit, and Hafey. The winning team's pitchers are McMullen, High, and Sax!

Jerry: And the score comes out to 8–7.

⟨pause⟩

Jim: I'm getting a funny feeling.

That worked out pretty nicely.

Jerry: Yeah.

It's almost too good to be true, isn't it? I wonder if Nancy has something up her sleeve.

Jim: Well.

⟨pause⟩

Well anyhow we have the pitchers! Maybe that will help us sort out the position players.

Jerry: Maybe.

But that's a tall order. There are all those guys. And this is just some random minor league game.

> Nonetheless, there was further progress. See what you can do. Or turn the page. Or get something to eat.

10.3   TEAM SPIRIT

Jerry: Here's an easy one: Watkins and Trosky were hit by pitchers Hafey and Proudhomme—both losing team pitchers—so Watkins and Trosky were on the winning team. Trosky is a second baseman. There's only one other second baseman, Arnovich, so he must play for the losers.

Jim: We haven't applied our favorite equation yet. For the two teams together, we have PA = 66 outs + 9 LOB + 15 runs = 90.

Jerry: Ah! Several players have 6 at-bats. That means one team has more than 45 plate appearances. So the other team must have fewer.

Jim: Only one team can have guys with 6 at-bats. So Gustine, Gruhe, and Melillo—they all have 6 ab—are all on the same team. And that team's PA has to be at least $3 \times 6 + 6 \times 5 = 48$. But which team is it? Is it the winners or the losers?

Jerry: The Hornets have 7 LOB, 33 outs, and either 7 or 8 runs, for 47 or 48 PA. That's the team with more plate appearances. But as you say, they have at least 48. That means they have exactly 48. And so they have exactly 8 runs. They're the winners! The Muskrats lost.

Since Gustine, Gruhe, and Melillo are Hornets, the other players at their field positions, Sisti, York, and Keesey, are Muskrats.

Jim: The Hornets won. We already showed that the visitors won. So the Hornets are the visitors. I'll make a chart. I'll put the Hornets on the left as the visitors.

Let's see what we have now.

⟨pause⟩

Jim: Okay, this is it so far:

Hornets	ab^+	r	h	bi
Gruhe cf	6	0	0	0
Melillo lf	6	1	1	0
Gustine 3b	6	1	1	1
Watkins ph,rf	1	0	0	1
Trosky 2b	5	2	2	0
Sax p	1	0	0	0
High p	1	0	0	0
McMullen p	2	2	2	0
Totals:	28	6	6	2

Muskrats	ab^+	r	h	bi
Keesey lf	5	1	1	1
Sisti 3b	5	0	0	0
York cf	3	4	2	0
Arnovich 2b	3	1	0	0
Hafey p	0	0	0	0
Douthit p	1	0	1	0
Brickell p	0	0	0	0
Suhr p	2	0	1	0
Cuccinelli p	1	1	1	1
Proudhomme p	0	0	0	0
Totals:	20	7	6	2

The "ab^+" there—I made that column at-bats plus HBP plus S and SF (plate appearances that don't count as at-bats). It's PA minus walks. I placed Gustine, Gruhe, and Melillo at the top since any players with fewer ab^+ have to bat later. Similarly, I placed Keesey and Sisti at the top of the Muskrats.

Jerry: The Muskrats now have all their runs. So we have to put Clancy and Rabbitt, who score runs, on the Hornets, and consequently Loftus and Batzberger, who play the same positions on the field, on the Muskrats.

Jim: That's everybody except the catchers, the right fielders, and Heineman, subbing at short.

⟨pause⟩

Jim: Jerry?

⟨pause⟩

Jim: Jerry?

Jerry: Yeah.

Yeah, I know. This is too easy. We're getting too far. You think there's no way Nancy could have just hit on the one box score out of a hundred that you could cut up and have a chance of piecing back together.

So maybe she just got lucky. You don't know my sister. She's smart alright — she could figure out our puzzles if she took the time and trouble. But she has no real interest in baseball. She only follows it enough to have conversations with our Dad, who's a big Cubs fan. She wouldn't have bothered to do the painstaking work of searching for promising box scores that make good puzzles, like we do.

Jim: She's the only person who's shown any interest in our puzzles so far. But have it your way.

⟨pause⟩

OK, where were we?

Jerry: We knew the total runs for each team from the pitching stats, and that told us how many runs each team's batters scored, and that enabled us to place most of the batters.

Jim: What if we try the same approach using the total hits and rbis for both teams?

Okay. From the pitchers' stats, . . .

Muskrats	ip	h	r	er	bb	so
Brickell	2	1	1	1	0	0
Cucc.	3	2	1	1	0	1
Douthit	1	2	1	1	0	1
Hafey (L)	2	2	1	1	0	2
Proud.	.2	2	0	0	0	2
Suhr	2.1	6	4	4	0	0

Hornets	ip	h	r	er	bb	so
High	4	5	2	2	0	0
McMul.	4	6	5	5	3	1
Sax (W)	3	1	0	0	0	3

we can see that the Hornets' pitchers give up 12 hits, so the Muskrats' batters must have 12 hits. In the updated chart below, with Loftus and Batzberger, the Muskrats have just 9 hits.

Hornets	ab$^+$	r	h	bi
Gruhe cf	6	0	0	0
Melillo lf	6	1	1	0
Gustine 3b	6	1	1	1
Watkins ph,rf	1	0	0	1
Trosky 2b	5	2	2	0
Sax p	1	0	0	0
High p	1	0	0	0
McMullen p	2	2	2	0
Rabbitt ss	4	1	2	0
Clancy 1b	5	1	3	2
Totals:	37	8	11	4

Muskrats	ab$^+$	r	h	bi
Keesey lf	5	1	1	1
Sisti 3b	5	0	0	0
York cf	3	4	2	0
Arnovich 2b	3	1	0	0
Hafey p	0	0	0	0
Douthit p	1	0	1	0
Brickell p	0	0	0	0
Suhr p	2	0	1	0
Cuccinelli p	1	1	1	1
Proudhomme p	0	0	0	0
Loftus ss	4	0	1	3
Batzberger 1b	5	0	2	0
Totals:	29	7	9	5

We need 3 more hits for the Muskrats and 4 more hits for the Hornets from the guys that are left:

	PA	r	h	bi
Worthington rf	4	0	2	2
Davis rf	5	0	1	1
O'Doul c	5	0	0	0
Therenow c	4	0	2	1
Heineman ph,ss	2	0	2	2

Jerry: Each team has to have a catcher and right fielder. One way to work this is for Therenow and Davis to play for the Muskrats, with O'Doul, Worthington, and Heineman on the the Hornets.

Jim: There's one other way. Put O'Doul, Davis, and Heineman on the Muskrats, then put Therenow and Worthington on the Hornets.

	PA	r	h	bi		PA	r	h	bi
Worthington rf	4	0	2	2	Worthington rf	4	0	2	2
Davis rf	5	0	1	1	Davis rf	5	0	1	1
O'Doul c	5	0	0	0	O'Doul c	5	0	0	0
Therenow c	4	0	2	1	Therenow c	4	0	2	1
Heineman ph,ss	2	0	2	2	Heineman ph,ss	2	0	2	2
			3/4	2/4				3/4	3/3

(green for the Muskrats, red for the Hornets)

Jerry: The two ways wind up giving different numbers of rbis. The second way gives the Muskrats 3 more rbis for a total of 8. That's too many since the Muskrats score only 7 runs. So it has to be the first way. That rounds out both teams. That's the lot.

Hornets	PA	r	h	bi
Gruhe cf	6	0	0	0
Melillo lf	6	1	1	0
Gustine 3b	6	1	1	1
Watkins ph,rf	1	0	0	1
Trosky 2b	5	2	2	0
Clancy 1b	5	1	3	2
O'Doul c	5	0	0	0
Worthington rf	4	0	2	2
Rabbitt ss	4	1	2	0
Heineman ph,ss	2	0	2	2
Sax p	1	0	0	0
High p	1	0	0	0
McMullen p	2	2	2	0
Totals:	48	8	15	8

Muskrats	ab+	r	h	bi
Keesey lf	5	1	1	1
Sisti 3b	5	0	0	0
Batzberger ss	5	0	2	0
Loftus 1b	5	0	1	3
Davis rf	5	0	1	1
Therenow c	4	0	2	1
York cf	3	4	2	0
Arnovich 2b	3	1	0	0
Hafey p	0	0	0	0
Douthit p	1	0	1	0
Brickell p	0	0	0	0
Suhr p	2	0	1	0
Cuccinelli p	1	1	1	1
Proudhomme p	0	0	0	0
Totals:	39	7	12	7

Jerry: The Hornets didn't have any walks, so we're sure about each Hornet player's PA. The Muskrats have 3 walks from McMullen. Until we know who got the walks, we only know their ab$^+$.

⟨long pause⟩

Jim: It's time you admitted it, Jerry. This is not a box score. It's a puzzle. Your sister composed it. She's a master puzzler. This beats anything we've done.

Jerry: It can't be a puzzle!

A puzzle of this scope is unimaginable. The closest thing we've done is Scraps[1] which I composed—and your extension, Leftovers.[2] Constructing Scraps, getting all the pieces to fit together, to talk to each other, to do what they had to do . . .

It took me weeks of intense concentration.

I lived on coffee and granola.

This may take some time to work out. There have been three installments so far. But we're just getting started.

In the next stage, our heroes look at three players with puzzling statistics: Proudhomme, Watkins, and Heineman. See what you can deduce about them.

Or don't. Just read on.

But maybe you need a rest? Take your time! We can wait.

---

[1]Chapter 5
[2]Chapter 7

10.4 THREE ODD MEN

⟨The Next Day⟩

Jim: ⎫ ⎧ Jerry, I'm embarrassed to admit that actually I-
Jerry: ⎭ ⎩ Jim, I have to tell you this, listen, I-

*Jerry, I was **wrong!** You were right. It's a real box score. I looked up the names and they're honest-to-God ballplayers—from about **a hundred years ago!** They weren't great. One guy only played five games of major league ball. They must have played together on a small-town team somewhere. **And the box score is obviously the real thing!***

Hey have you been listening to me?

*Jim, I was **dead wrong**. You were right. It's got to be a puzzle. I just couldn't accept that Nancy made a puzzle so complex. **It's better than anything I could do.** But, but look! Everything is falling neatly into place! Some players like Douthit (Douthit?) were clearly made up! The teams are a joke. And the box score—**the box score is a total fake!***

Have you heard anything I've said?

⟨A minute of silence⟩

Jerry: You know, some of these players present interesting situations, the pitcher Proudhomme for example.

Jim: Oh? What about him?

Proudhomme .... ⅔ 2 0 0 0 2

Jerry: Proudhomme pitches a fragment of an inning—2 outs, both of them strikeouts. He also lets at least 3 runners get on base—2 with hits and he hits one guy with a pitch. But he gives up no runs. It seems that he must leave all 3 runners on base. Is that right?

Jim: Maybe there are more than 3 runners. What if the catcher dropped a third strike?

Jerry: A dropped third strike wouldn't change things. Proudhomme still faces 5 batters—2 hits, 2 strikeouts, and 1 guy hit by a pitch. But no run is charged to him, so no matter what, when he leaves the game, the bases will be loaded. And none them will score.

Jim: But I think you're assuming that Proudhomme pitches in only one inning. Couldn't he pitch parts of two successive innings?

Jerry: Oh!

Let's see ... He would have to end the first inning, getting the last out. If he got 1 out each in 2 innings, then there would have to be a pitcher before him and a pitcher after him with fractional innings pitched. But there aren't two such pitchers, so that's impossible.

And if Proudhomme got 2 outs in 1 inning and then pitched in the next inning getting no outs there would be a note about that in the box score, just like the note about Brickell pitching in the ninth. But there isn't a note.

Jim: Okay, he just pitches in 1 inning. If he starts the inning, his successor can't let any of the three guys on base score because in that case the run would be charged to Proudhomme, who gives up no runs. He might pick one off. That would leave two on base. Or the next batter might make an out. Then it's three left on base. In either case, if Proudhomme starts the inning, no runs are scored.

Jerry: If instead he starts pitching with 1 out, then there must be 3 LOB when he records his second out, ending the inning.

Let's remember this: 2 or 3 LOB when Proudhomme pitches.

Jim: Okay, that's Proudhomme. Now another guy that stands out is Watkins.

Watkins ph,rf 0 0 0 1

HBP–Watkins by Hafey,

The table of hitters we made shows he has only 1 plate appearance, which is when he pinch-hits for somebody and he gets hit by a pitch from Hafey. But he has an rbi! So he must be hit by a pitch with the bases loaded, forcing in a run.

When could that have happened?

Jerry: Umm… umm… umm…

Hafey gave up only one run, the run forced in by Watkins. And Hafey was the losing pitcher, so it must have been the winning run… Hey! The Hornets won 8–7 in 11 innings. The game must have been tied after 10 innings, so the winning run had to be scored in the eleventh inning. That's when Watkins was hit by the pitch.

Jim: That's a hell of a way to lose a game.

Jerry: Since the Hornets have 48 PA, their last 3 batters in the game are slots 1, 2, and 3, which are filled in some order by Gruhe, Melillo, and Gustine, the Hornets who have 6 at-bats each. Watkins can't pinch-hit in one of those slots because that would give the slot 7 PA—too many.

Instead, Watkins could have batted in slot 9 after the batters in slots 6, 7, and 8 got on base. Actually, that's how it must have been—7 players batting is the maximum for an inning with 1 run. So there were 3 LOB in the eleventh inning, and the run was driven in by Watkins and scored by the player in slot 6, who led off the inning.

Jim: Watkins must have pinch-hit for Worthington, and then taken over his place in right field in the bottom of the eleventh. Worthington's 4 PA and Watkins 1 PA would account for all 5 PA in slot 9 … But, wait, can we rule out a double switch, with Watkins pinch-hitting for someone else?

Jerry: Good question! But in that case, how would we account for Worthington's slot? He has only 4 PA, so he must be replaced by somebody with only 1 PA (the only 6 PA slots are taken by Gruhe, Melillo, and Gustine). If that player isn't Watkins, it must be Sax or High, the only other Hornets with just 1 PA (chart on p. 136). They're both pitchers.

Jim: The Hornets pitcher in the top of the eleventh is Sax, so it would have to be Sax. But if Watkins pinch-hits for Sax, there would be no pitcher left to pitch the bottom of the eleventh. So we can rule that out. Worthington had to be the original occupant of slot 9.

slot	who's in the slot
1	
2	Gruhe, Gustine, Melillo
3	(in some order)
4	
5	
6	
7	
8	
9	Worthington, then Watkins

Jerry: And then there's Heineman, a shortstop, who enters the game as a pinch hitter and stays in the game, replacing the starting shortstop, Rabbitt.

Heineman ph,ss 2 0 2 2

But he can't pinch-hit for Rabbitt. Rabbitt has 4 PA and Heineman has 2. That would give the slot a total of 6 PA, but the 6 PA slots are already taken. Heineman must pinch-hit for a batter with fewer than 4 PA. The only possibilities left are 2 pitchers, High (1 PA) and McMullen (2 PA). Heineman together with the pitchers have 5 PA; that's a slot.

Jim: This can only be in the top half of the ninth inning after High and McMullen (4 innings each) have finished pitching but before Sax comes in to pitch the last 3 innings. It's a double switch in the top half of the ninth with Heineman replacing the pitcher and the new pitcher, Sax, replacing Rabbitt.

Jim: Heineman has 2 at-bats and 2 rbis. When does he drive guys in?

Jerry: McMullen and High give up all 7 Muskrat runs in the first 8 innings, so the teams must be tied at 7 going into the tenth. The Hornets score only once after that, in the eleventh, and that run is driven in by Watkins. So Heineman must have had his 2 rbis in the ninth. Who pitched the ninth?

Jim: The loser, Hafey, pitched 2 innings, which had to be the tenth and eleventh. We know Brickell pitched the seventh and eighth. So the guy who pitched the ninth pitched only one inning. That has to be Douthit—no other Muskrat pitcher pitches just one inning. But Douthit gave up only 1 run, not 2!

Jerry: Ah! But remember, Brickell also pitched to 1 batter in the ninth (p. 130)! He gave up 1 run, so that must be the other run in the ninth.

By the way, that means the Hornets didn't score in Brickell's other innings, the seventh and eighth.

inning	1 2 3 4 5 6	7 8	9	10 11
Muskrat pitcher	???	B B	B/D	H H
Hornet runs	5	0 0	1 1	0 1

Muskrat pitching so far

Jerry: Have you noticed that the snippets of paper—the ends of the line scores that would tell us the final scores—they weren't included?

Jim: Hey yeah, that's right.

Jerry: Maybe, maybe—

⟨Pause.⟩

I'm going to call Nancy.

⟨dialing⟩

Jerry: Hi Nancy. We were just wondering. The snippets of paper that you sent didn't include the final scores. You know, the ends of the line scores ... ?

Nancy: Oh!

I guess the kids must have lost those. It was a real mess when they were cutting up the box score.

Say, you're not having trouble are you?

⟨A long pause.⟩

Jerry: No. Just wondered. Thanks.

⟨End of call.⟩

Jim: That didn't help.

Jerry: Yeah. Okay.

But so what?

We're having fun.

We're making progress on the puzzle or whatever it is.

Jim: She's your sister. But I don't think she's going to be helpful.

Jerry: That's for sure! And after that snide remark of hers, we're
        really going to have to rise to the challenge.

You might be able to put together the Hornets' pitching order.

Here's a hint: Pay attention to the batting statistics of the Hornets'
pitchers.

Whether that helps or not, the solution starts on the next page.

Jerry: For the Hornets pitching, all that's left is to figure out whether McMullen or High is the starter.

Jim: What's different about them?

Jerry: McMullen gives up 5 runs, and High only gives up 2. Can that help us?

Hornets' Pitching:

innings:	1 2 3 4	5 6 7 8	9 10 11	total
Hornet pitchers:	High or McM	McM or High	Sax	
Muskrat runs:	2 or 5	5 or 2	0	7

Jim: I don't see how. We have no idea when the Muskrats scored their runs. But what about their batting statistics. Could that somehow help us?

Jerry: McMullen has 2 PA and scores 2 runs, but High has only 1 PA and no runs. Is that helpful?

Jim: Actually, that's promising, because we just figured out that the Hornets didn't score in the seventh and eighth innings. Is it possible that McMullen could have scored his 2 runs in 2 successive innings, the fifth and sixth?

Jerry: He definitely couldn't score twice in the same inning — that requires 8 runs by the team. But why couldn't he score 1 run in the fifth inning and another 1 in the sixth?

Jim: But, Jerry, if McMullen is the second pitcher, I don't think he bats at all in the fifth. The Hornets are the visiting team, so they pitch in the bottom half of each inning. After the starting pitcher, High, (in this scenario) finishes pitching the bottom of the fourth inning, he's still in the game. He's the one who would bat in the top for the fifth, not McMullen.

Jerry: But how do you know McMullen didn't pinch-hit for him in the top of the fifth?

Jim: Because the box score would have told us. It would say: "McMullen ph, p," not just "McMullen p." Box scores always tell you when somebody enters as a pinch hitter. Just look at Heineman's line: "Heineman ph, ss." It's entirely analogous. In both cases a player pinch-hits for someone and replaces him at his fielding position. If McMullen had been a pinch hitter, it would have said so.

Jerry: OK, you've convinced me. McMullen couldn't bat in the fifth, and can't score 2 runs in the sixth, so he can't be the second pitcher. He must be the starter, and High his replacement. We've got the Hornets pitching settled.

Hornets' Pitching:

innings:	1 2 3 4	5 6 7 8	9 10 11	total
Hornet pitchers:	McMullen	High	Sax	
Muskrat runs:	5	2	0	7

10.5 THE MUSKRATS' PITCHING

Jerry: We already know the last three Muskrats pitchers: Brickell then Douthit then Hafey. We just have to order Suhr, Proudhomme, and Cuccinelli.

Jim: I've been thinking about this. It just seems like the pieces can fit together in so many different ways.

We could use a hint.

Jerry: Hint? Who's going to give us a hint?

If you want a hint, there's one on the next page.

And if you're in a hurry, the solution follows after that.

Hint:

The key is when players are left on base. Look at the number of LOB in the ninth and tenth innings. And the key for that is seeing where Heineman bats (and drives in 2 runs) in the ninth. Keep in mind the number of players the Hornets leave on base at various points in the game and note the fact that slot 6 leads off in the eleventh.

Who bats in slot 6 is also important.

It's all delicately balanced.

The Hornets had 2 or 3 LOB when Proudhomme pitched, which was sometime in the first 6 innings—that's what Jerry wanted remembered on page 140. The Hornets also had 3 LOB in the eleventh (p. 141). That leaves at most 2 LOB for the ninth and tenth innings.

Muskrats' Pitching:

innings:	1 2 3 4 5 6	7	8	9	10	11	totals
Muskrat pitchers:	???	B	B	B/D	H	H	
Hornet runs:	5	0	0	2	0	1	8
Hornet LOB:	≥2			≤2		3	7

The ninth and tenth innings have 2 runs, 6 outs, and 0–2 LOB for a total of 8–10 PA. Counting back from slot 6, which led off the eleventh,

slots		9 and 10	11
	1	5	
	2	4	
	3	3	
	4	2	
	5	10 1	
	6	9	●
	7	8	●
	8	7	●
	9	6	●

the ninth inning begins with slot 5, 6, or 7. But Heineman, who doesn't homer, can't get 2 rbis unless he bats third in the inning or later, that is, in slot 7, 8, 9, 1, 2, or 3. But it's not slot 9, which is taken by Worthington and later by Watkins. He also can't be in slots 1, 2, or 3 which are taken by the guys with 6 PA, Gruhe, Melillo, and Gustine. That leaves 7 or 8.

Now suppose Heineman was in slot 8. Then in the eleventh he would have batted with 2 runners on base. He has 2 hits, a double and a triple. Either one would have driven in a run—but the winning run must be driven in by Watkins' HBP.

So Heineman must be in slot 7 and drive in the players in slots 5 and 6 (the first 2 batters of the ninth) for his 2 rbis.

And since slot 5 leads off the ninth, the chart shows that there are 10 batters in the ninth and tenth combined, and thus 2 LOB in these innings.

Since Heineman pinch-hit for pitchers McMullen and High, they must also bat in slot 7. And by the way, since the player in slot 6 scores in both the ninth and eleventh innings, he must be Trosky, the only Hornets player remaining with 2 runs.

slot	who's in the slot
1	
2	Gruhe, Gustine, Melillo
3	(in some order)
4	
5	
6	Trosky
7	High then McMullen *or* McMullen then High, *then* Heineman
8	
9	Worthington, then Watkins

We have shown that there are 2 LOB in the ninth and tenth innings. We already knew there were 3 LOB in the eleventh. That leaves exactly 2 LOB for the inning when Proudhomme pitched, whenever that was.

We can now pin down the Muskrats' pitching order. Proudhomme can't enter in the middle of an inning, because as we saw on page 140 that would require 3 LOB in that inning. So Proudhomme (with two-thirds of an inning) starts an inning and must be followed by Suhr, who has 2 and a third innings. Proudhomme faces 5 batters (2 outs, 3 on base) and he hits Trosky in slot 6, so he wasn't the first pitcher in the game. And since Proudhomme is charged with no runs, none were scored in his inning. We now have the pitching order:

Cuccinelli—Proudhomme—Suhr—Brickell—Douthit—Hafey.

Muskrats' Pitching:

innings:	1 2 3	4	5 6	7 8	9	10 11	totals
Muskrat pitchers:	CCC	P/S	S S	B B	B/D	H H	
Hornet runs:	1	0	4	0 0	2	0 1	8
Hornet LOB:	0	2	0 0	0 0	2	3	7

10.6 THE HORNETS' BATTING

Jim: I'm going to make a chart that summarizes what we can conclude about which Hornet bats in which slot and in which innings the various slots bat, based upon the LOB information we've already deduced.

We know that McMullen as a batter for the Hornets scored 2 runs in his 2 plate appearances in slot 7, so I'll put that in the chart. In the first 3 innings there are no LOB and Cuccinelli gives up only McMullen's run, so the first 6 batters must make outs. So I'll put just 3 batters and no runs in each of the first 2 innings. Then McMullen leads off the third and eventually scores.

Jerry: We also know there are exactly 5 plate appearances in the fourth, so we know McMullen's second run is in the fifth.

Jim: Right. I've got that. But we don't know how the 4 runs scored against Suhr are split between innings 5 and 6. One run is McMullen's in the fifth inning but the other 3 runs might be in either inning. So I'll put three dots on the line between those innings.

Jerry: *And* we don't know where in the ninth and tenth innings the 2 left on base go. We have work to do!

Hornets	1	2	3	4	5	6	7	8	9	10	11
Gustine Melillo Gruhe	•		•		•		•		•		•
	•			•		•		•		•	•
	•			•		•		•		•	•
		•		•		•		•		•	
		•		•		•		•◇	•		
Trosky		•		•		•		•◇		◇•	
McM/Hi/Hman			◇•		◇•		•		•		•
				•		•		•		•	•
Worth/Wat				•		•		•		•	•
Runs	0		1	0	4		0	0	2	0	1
LOB	0	0	0	2	0	0	0	0	2		3

Jerry: You know, I've been avoiding looking at those snippets of numbers that we think are the cut-up line scores. But now that we've got the pitchers in order, maybe it's time to see if we can piece together those line scores.

Jim: Yeah, like that little block of 13—that can't possibly be 13 runs in a single inning! It must be that one of the teams had 1 run and then 3 runs in successive innings. Same for the block of 21. Now that we know for each pitcher when he pitches and how many runs he gives up, maybe we can figure out when the remaining runs are scored.

Jerry: And I checked that those loose numbers do add up to 15, so they account for all 15 runs in the game. And there are 22 digits in all, just right for 11 innings (with missing final scores).

Jim: Let's see what else we can figure out about when the remaining runs scored and who batted in each slot.

When indeed, and where do all the Hornets players bat? There's a hint on the next page. And the solution follows.

Hint:

To get started, deduce which team has the 13 runs block and which team has the 21 runs block.

Start with the line score for the Hornets.
Looking at McMullen's runs will help.

The tricky part will be innings 5 and 6.
But that should help you figure out slots 1, 2, and 3.

We now have all the runs for the Hornets pinned down except for the fifth and sixth innings (x and y):

$$001\ 0xy\ 002\ 01$$

We just showed that McMullen scores in the fifth, so x is at least 1. xy could be 13, 22, or 31, since High gave up 4 runs. But xy can't be 40, because the snippets of runs information includes no number 4.

Now there has to be a block of 13 and a block of 21 in the two teams' line scores. The 21 block doesn't belong to the Hornets, so it must belong to the Muskrats. What about 13 block? If both blocks are in the Muskrats' line score, then 21 and 13 would make up all of the Muskrats' scoring. But the Hornets' pitcher McMullen gives up 5 runs and then High gives up 2 runs. Neither 13 21 nor 21 13 has an initial string of 5 runs. So 13 must be in the Hornets' line score. And that means that xy is 13. So now we have the Hornets' line score,

$$001\ 013\ 002\ 01$$

and an improved scorecard:

Hornets	1	2	3	4	5	6	7	8	9	10	11
	•		•	•		•			•		•
Gustine	•		•		•		•	•		•	
Melillo	•		•		•		•		•	•	
Gruhe		•		•		•		•		•	
		•		•		•		◈	•		
	•		•		•			◈		◈	
Trosky		◈		◈	•				•		•
McM/Hi/Hman			•		•		•		•		•
			•		•		•		•		•
Worth/Wat			•		•		•		•		•
Runs	0	0	1	0	1	3	0	0	2	0	1
LOB	0	0	0	2	0	0	0	0	2		3

Back to the Muskrats lineup. It will help to summarize what we know about the players:

In slots 1–3, we have, in some order:

	PA	r	h	bi	special
Gustine 3b	6	1	1	1	
Melillo lf	6	1	1	0	3B
Gruhe cf	6	0	0	0	

For slots 4, 5, and 8, we have, in some order the following possible players:

	PA	r	h	bi	special
Rabbitt ss	4	1	2	0	2B
O'Doul c	5	0	0	0	
Clancy 1b	5	1	3	2	

We've already assigned the other players to slots, including

	PA	r	h	bi	special
6. Trosky 2b	5	2	2	0	2B, HBP (Proudhomme)
7. McMullen p	2	2	2	0	2B(2)
7. High p	1	0	0	0	
7. Heineman ph, ss	2	0	2	2	2B, 3B
9. Worthington rf	4	0	2	2	
9. Watkins ph, rf	1	0	0	1	HBP (Haley)

From the improved scorecard we see that all the runs (indicated by diamonds) are accounted for by slots 5, 6, and 7—except for the 3 runs in the sixth inning. Slot 8 doesn't bat in the sixth inning, so it has no runs. There are only 3 candidates listed above as possible for slot 8, and the only one with no runs is O'Doul. So he must be in slot 8.

Similarly, slot 1 doesn't bat in the sixth inning, so it can't score a run. Since Gustine and Melillo each do score a run (which must be in the sixth inning), that means Gruhe must be in slot 1.

Since O'Doul in slot 8 and Gruhe in slot 1 have no rbis and McMullen doesn't have a homer, Worthington must have driven in McMullen in both the third and fifth innings, accounting for his 2 rbis. Now all rbis are accounted for except for those in the sixth inning, as in this much-improved scorecard:

Hornets	1	2	3	4	5	6	7	8	9	10	11
Gruhe	•		•		•		•		•		•
Melillo	•			•		•		•	•		•
Gustine	•			•		•		•		•	•
			•		•		•		•		
			•		•		•		•		
Trosky			•		•		•		•		◇•
McM/Hi/Hman			◇•		◇•				•		•
O'Doul				•		•		•		•		•
Worth/Wat				•		•		•		•		•
Runs	0	0	1	0	1	3	0	0	2	0	1
LOB	0	0	0	2	0	0	0	0	2		3

Slot 2 leads off the sixth inning and neither Gustine nor Mellilo has a homer, so slot 2 can't have an rbi in this inning, and since the rbis in all other innings are all accounted for in other slots, slot 2 can't have any rbi at all. But Gustine does have an rbi. So Melillo must bat in slot 2 and Gustine in slot 3. Gustine must drive in Melillo in the sixth.

This leaves only Clancy and Rabbitt to fill slots 4 and 5. Clancy has 2 rbis but no homer, so he can't be in slot 4, where he could only drive in one player, Gustine. So Rabbitt is in slot 4 and Gustine in slot 5. Clancy drives in both Gustine and Rabbitt.

Hornets	1	2	3	4	5	6	7	8	9	10	11
Gruhe	•		•		•		•		•		•
Melillo	•			•		◇•		•		•	•
Gustine	•			•		◇•		•		•	•
Rabbitt		•		•		◇•		•		•	
Clancy		•		•		•			◇•	•	
Trosky		•		•		•		◇•		◇•	
McM/Hi/Hman			◇•		◇•				•		•
O'Doul			•		•		•		•		•
Worth/Wat			•		•		•		•		•
Runs	0	0	1	0	1	3	0	0	2	0	1
LOB	0	0	0	2	0	0	0	0	2		3

Melitto leads off the sixth with a triple, since it's the only way he can reach base. He has no other hits, walks or HBP; there are no errors; and there are no WP or PB in the game to allow anybody to reach base on a strikeout. Gustine could then drive him in with his single or else get on base via a fielder's choice.

The final task for the Hornets' scorecard is to decide whether slots 1 and 2 bat in the ninth or tenth innings. The key is Haley. We have seen that he gives up a run in the eleventh inning when slots 6–9 all get on base. Filling in the players' names, Trosky leads off. Like Melillo in the sixth inning, the only way he can reach base is with a hit. Then Heineman gets a hit, too (it must be his double, or else he would drive in Trosky instead of Watkins, who drove him in with his HBP). That accounts for both of the hits given up by Haley. There is no longer any way anybody can reach base in the tenth inning against Haley. So there can only be 3 batters in the tenth, slots 3, 4, and 5, and they must all make immediate outs. Slots 1 and 2, Gruhe and Melillo, bat in the ninth.

Here's a pretty complete chart, together with all the rbis. It incorporates Trosky's HBP by Proudhomme in the fourth inning and Melitto's triple. The switches and double switches involving slots 4, 7, and 9 are marked by gray lines. Note that since McMullen leads off the fifth, Trosky must have been the last plate appearance for the Hornets in the fourth. But Suhr got the last out, so he must have pitched to McMullen, and a base runner must have been put out to end the inning before that plate appearance was completed.

We also allocate all the hits for each player into particular innings, even if we can't always prove that they occurred in those particular innings. All events shown in the chart or in notes below it that we have not proved are highlighted in blue. The purpose of including such events is to help show that it is really possible to have a game that is fully consistent with all information in the box score, not that the game must have transpired exactly this way.

Hornets	1	2	3	4	5	6	7	8	9	10	11	ab	R	H	rbi
Gruhe cf	•		•		•		•		•		•	6	0	0	0
Melillo lf	•			•	3		•	•			•	6	1	1	0
Gustine 3b	•			•	1		•		•	•		6	1	1	1
Rabbitt ss		•		1	2		•					4	1	2	0
Sax p									•			1	0	0	0
Clancy 1b		•		1		1		1	•			5	1	3	2
Trosky 2b		•		hbp		•		2		1		4	2	2	0
McMullen p			2		2							2	2	2	0
High p						•						1	0	0	0
Hein'm'n ph,ss									3	2		2	0	2	2
O'Doul c				•		•		•		•		5	0	0	0
Worthington rf				1		1		•		•		4	0	2	2
Watkins ph,rf											hbp	0	0	0	1
TOTALS (R/H)			1/2	/2	1/2	3/4			2/3		1/2	46	8	15	8

Note: Suhr picks Clancy off second base to end the Hornets' half of the fourth inning.

10.7 THE MUSKRATS' BATTING

Jim: We're in the home stretch, Jerry! All we have to do is figure out how the Muskrats scored their runs.

Jerry: Believe it or not, we still haven't filled out their line score. But we do know the pieces.

All remaining snippets of runs belong to the Muskrats: a 2-inning snippet 21, and individual snippets of 3 and 1. The snippets must be arranged in such a way that McMullen gives up 5 runs in the first 4 innings and High gives up 2 in the next 4 innings.

The 3 snippet can't be in High's region, so it's in McMullen's. The 21 can't be entirely in High's territory, or High would have 3 runs. It can't be in the eighth and ninth innings, because the ninth has no runs. The 21 block can't be all in McMullen's region, because that would give him 6 runs. It must be split: with the 2 in the fourth inning and the 1 in the fifth inning. High must have the two 1-run innings.

Muskrats:

inning:	1 2 3	4	5	6 7 8	9 10 11
runs:	3	2	1	1	0 0 0

Jim: Looking at York should be helpful. He scores 4 runs. These have to be in separate innings because if a guy scores twice in an inning, that's at least 11 plate appearances, meaning at least 8 runs. So York scores in each of the scoring innings: 3, 2, 1, and 1.

Jerry: Also helpful is the fact that the Muskrats left only 2 men on base. Maybe with this we can figure out how many guys bat in every inning.

Jim: We have to start with innings 4 and 5, since York must score in both. At the very least, the 2 innings must have 11 PA (York bats twice—10 PA—but a scorer can't be the last batter—11 PA). The 2 innings have 6 outs and 3 runs. Both LOBs are needed to get up to 11 PA. So the innings have exactly 11 PA.

Innings 1–3 have 3 runs, 9 outs, and no LOB, for 12 PA. The fourth inning starts with slot 4. Since the fourth and fifth together have 11 PA, the fifth ends 11 PA later with slot 5. And York has got to be in slot 4 to score twice.

Jerry: Great! But we still don't know where the dots go for innings 1–3 and 6–8. We know the rest, though, since they have no runs or LOB. I'll start the scorecard. Give me a minute.

⟨A minute⟩

	1	2	3	4	5	6	7	8	9	10	11
				•						•	
					•					•	
					•					•	
York				◇•	✕ •◇						•
				•	•						•
				•							•
				•				•			
				•				•			
					•			•			
RUNS		3		2	1		1		0	0	0
LOB	0	0	0	2	0	0	0	0	0	0	0

Jim: Hey, York scores the only run of the fifth, so he must be one of the first 3 batters of the inning, hence those dots between the fourth and fifth innings go to the fourth inning.

> Okay, your turn!
> But the solution follows.

The Solution

In the list of the Muskrats' players (p. 136), we identified 5 of the 6 players who have 5 PA. Since York is in slot 4, he must also have 5 PA, and thus must have 2 walks in addition to his 3 AB. Slots 7–9, with 4 PA each, must be filled with Arnovich, Therenow, and the pitchers. There are no substitutions.

The only Hornets pitcher delivering walks is McMullen, the first pitcher, so York must be walked in his first 2 at-bats. Arnovich must have the third walk given up by McMullen, to bring his PA up to 4.

Muskrats	ab+	r	h	bi
Keesey lf	5	1	1	1
Sisti 3b	5	0	0	0
Batzberger ss	5	0	2	0
Loftus 1b	5	0	1	3
Davis rf	5	0	1	1
York cf	3	4	2	0
Therenow c	4	0	2	1
Arnovich 2b	3	1	0	0
Hafey p	0	0	0	0
Douthit p	1	0	1	0
Brickell p	0	0	0	0
Suhr p	2	0	1	0
Cuccinelli p	1	1	1	1
Proudhomme p	0	0	0	0
Totals:	39	7	12	7

		1	2	3	4	5	6	7	8	9	10	11
Keesey					•					•		
Sisti						•				•		
Loftus						•				•		
Davis / York					◇✕◇•						•	
Batzberger					•	•					•	
					•						•	
Arnovich					•			•				
Therenow					•			•				
pitchers					•			•				
RUNS		3		2	1		1		0	0	0	
LOB	0	0	0	2	0	0	0	0	0	0	

The inning in which the Muskrats score 3 runs has 6 batters (3 runs, 3 outs, and no LOB). Which inning is it? Not the first, because the first 6 batters in the lineup (the ones with 5 PA) include only two who score any runs. And not the third, because then the 6 batters wouldn't include York, who leads off the fourth. So it must be the second inning.

And in which inning (of the sixth through the eighth) did the Muskrats score a single run? It must be York who scored it for his fourth run. Since nobody was left on base in those innings, everybody not in York's slot (#4) made outs. From the scorecard we see that the sixth inning begins with slot 6. So the sixth inning had outs by the batters in slots 6–8, the seventh had outs from slots 9, 1, and 2, and in the eighth inning slots 3–6 batted, including York, who scored. That nails down the placement of PAs:

		1	2	3	4	5	6	7	8	9	10	11
Keesey		•		•	•		•				•	
Sisti		•		•		•	•				•	
Loftus		•		•		•		•			•	
Davis	York		W		W	•		•				•
			•		•	•		•				•
Batzberger			•		•		•	•				•
Arnovich			•		•		•		•			
Therenow			•		•		•		•			
pitchers			•		•			•	•			
RUNS		0	3	0	2	1	0	0	1	0	0	0
LOB		0	0	0	2	0	0	0	0	0	0	0

From the scorecard we can see that slots 2 and 3 can't have any rbis. They only bat in 2 innings in which the Muskrats score, the fifth and eighth. In those innings only York scores, and slots 2 and 3 bat before him, and so can't drive him in. So slots 2 and 3 have no rbis in the whole game. Only 2 of the first 6 batters (those with 5 PA) have no rbi, Sisti and Batzberger, so they are in slots 2 and 3 (in some order).

The scorers other than York (Keesey, Arnovich, and Cuccinelli), must score in the second or fourth inning, the only innings with runs not already allocated to York.

Keesey, with 5 PA, is somewhere in slots 1–6. He has a homer and only one rbi, so it's a solo homer. He can't hit a homer in slot 1 since slot 1 doesn't bat in the second inning and bats last in the fourth. Slots 2, 3, and 4 are already taken. Could Keesey be in slot 5? But York in slot 4 scores in both the second and fourth innings, and has no homer, so he would be on base when Keesey, the next batter, homered, and that would give Keesey 2 rbi — impossible. So Keesey is in slot 6.

In the inning when Keesey homers (the second or fourth), it must be the slot 5 batter who drives in York, or else Keesey would have a 2-run homer. The only batter who can drive in York in the fifth inning is also the slot 5 batter. Same for the eighth inning: Keesey in slot 6 has no rbis left, so slot 5 must drive in York. So Loftus, the only Muskrat with 3 rbi, is in slot 5.

This leaves only one batter with 5 PA who can possibly be in slot 1, Davis. Davis has an rbi batting last in the fourth inning. He can't drive in York, who is six steps away. He can't drive in Keesey or Cuccinelli, who drive themselves in with homers. The only other Muskrat with any runs is Arnovich, so that's who he drives in.

		1	2	3	4	5	6	7	8	9	10	11
	Davis	•		•	•			•			•	
Sisti		•		•		•		•			•	
Batzberger		•		•		•			•		•	
	York	W		W	•			•				•
	Loftus		•		•	•			•			•
	Keesey		•		•		•		•			•
Arnovich			•		•		•			•		
Therenow			•		•		•			•		
pitchers			•		•			•		•		
	RUNS	0	3	0	2	1	0	1	0	0	0	0
	LOB	0	0	0	2	0	0	0	0	0	0	0

Arnovich has only 4 PA, so he's somewhere in slots 7–9. But which one? There are 2 runs in the fourth inning, #4 York and Arnovich, so by the Gap Rule all intervening players must be out—that's at most 2 intervening players. Arnovich must be in slot 7.

With all other runs accounted for, Keesey's and Cuccinelli's homers must be in the second inning. Cuccinelli can't homer as the last batter in the inning, so he must be in slot 8. That leaves Therenow for slot 9. His rbi can't be in the first inning when he bats after Cuccinelli's homer with the bases empty. So his rbi must be in the fourth, driving in York.

	1	2	3	4	5	6	7	8	9	10	11
Davis	•		•	•			•			•	
Sisti		•		•	•		•			•	
Batzberger		•		•	•			•		•	
York		⟨w⟩		⟨w⟩	•			⟨•⟩			•
Loftus		•		•	•			•-			•
Keesey	⟨HR⟩		•		•		•				•
Arnovich		•		⟨•⟩			•		•		
Cuccinelli	⟨HR⟩		•		•		•				
Therenow		•		•			•		•		
RUNS	0	3	0	2	1	0	1	0	0	0	0
LOB	0	0	0	2	0	0	0	0	0	0	0

Davis grounds into a double play. That can only happen in the seventh inning—in the fourth where he drives in Arnovich, there are already 2 outs (slots 5 and 6). Sisti grounds into a double play twice. There must be a runner on base each time. If Sisti is in slot 2, this isn't possible, for the following reasons: Davis leads off in the first, third, and tenth innings, but has only one hit and there is no error, WP, or PB that would let him get on base any other way. In the seventh Davis is out on a double play. So Sisti must be in slot 3, and Batzberger in slot 2.

The scorecard below incorporates what we can prove. Included as well are possible placements of walks, hits, etc., that are consistent with the data and the solution. As before, everything in blue is extra, that is, we don't have proof; they are there to show one way the game could have gone, consistent with all the data (the snippets).

The Muskrats	1	2	3	4	5	6	7	8	9	10	11	ab	R	H	rbi
Davis rf	•	•	•	1			DP			•		5	0	1	1
Batzberger ss	•	•	1		1	•				•		5	0	2	0
Sisti 3b	•	•	DP		DP			•		•		5	0	0	0
York cf		W		W	2		3				•	3	4	2	0
Loftus 1b		SF		•	1			•			•	4	0	1	3
Keesey lf		HR		•	•		•				•	5	1	1	1
Arnovich 2b		•		W	•			•				3	1	0	0
Cuccinelli p		HR										1	1	1	1
Proudhomme p												0	0	0	0
Suhr p				1	•							2	0	1	0
Brickell p												0	0	0	0
Douthit p								2				1	0	1	0
Hafey p												0	0	0	0
Therenow		•		1		1		•				4	0	2	1
TOTALS	R	3	1	2	1			1				38	7	12	7
	H	2		3	3		1	1	1						

Notes:

In the second inning, York steals second base and goes to third on McMullen's balk.

Suhr is thrown out at third base to end the fourth inning.

In the fifth, after York scores, Loftus is thrown out at second base.

In the ninth, Douthit is picked off second base.

⟨Later . . .⟩

Jerry: Well, Jim. We did it. We worked out far more than we thought possible,. How long did this take us, a month? It seems like that, anyway.

Jim: Three weeks.

Jerry: I'm going to email Nancy our final scorecards. That's the whole story. She can read between the lines.

⟨Just a little later, Nancy calls.⟩

Nancy: Congratulations guys! You are super!

Was it fun?

Jim: It was! What a trip!

Jerry: It was an incredible challenge. You really put us through the wringer. You ...

⟨pause⟩

Nancy, why didn't you tell us it was a puzzle? That ...

We agonized over that!

Nancy: That was part of the puzzle. I worked hard to make it look good.

But I really thought you would quickly see through my story. I just wanted to add something special.

Jim: It was special all right. All those clippings looked genuine. And I tracked down the players you used.—obscure ball players from the 1920's. You really had us going!

Nancy: It took me a long time to get the right paper and the right fonts and to dig up those guys. But I goofed on the box score format. Old box scores look different. Didn't that tip you off?

Jerry: It did! And ... and ... it didn't!

Nancy: If I had told you it was a puzzle—well, you might just have put it away. Or you'd have bugged me for hints. This way, I thought I might hook you.

Jim: Hooked!

⟨long pause⟩

Jerry: Well, all I can say, Nancy, is that I'm proud of you. Mighty proud to have a sister who put together a puzzle that I couldn't even imagine. And you never showed much interest in baseball before!

You're my sister. But you're no longer my "little sister"!

Hints

Hints for <u>Lucca</u> (p. 58)

Jim: This is nuts. 58 runs! These games are out of control.

Jerry: It helps to realize that since the losing team plays a steady game without getting any runs, it must have the same number of LOB in every inning. So if Brooklyn is the losing team, it must play 11 innings to get its 11 LOB. And the score must be 0-0 after the tenth inning. But then there would be no way Lucca could get 15 LOB in 11 innings. So Lucca must be the loser. It has to play 5 innings. Are you with me so far?

Jim: No.

Jerry: Then apply our favorite equation, PA = Runs + Outs + LOB, and show it's impossible for Brooklyn to score 58 runs.

Jim: Of course it won't. That would be ridiculous.

Jerry: But 57 runs and 59 runs are possible.

Jim: **! ! ! !**

Solution on p. 188

Hints for <u>Ancient Rivalry</u> (p. 59)

Jim: Jerry, I need help.

Jerry: Yeah, this isn't easy. Try answering these questions in order:

> In what innings does Yastrzemski score?
> When does Bressoud drive in runs?
> Who leads off the fourth inning?
> When do Bressoud and Yastrzemski hit homers?
> When did Schilling score his run?
> Who drove in whom in the fifth inning?

> If you can do that, you're pretty much done.

Jim: And can I figure all this out without using the fact that the box score gives the homers in the order in which they came in the game?

Jerry: Um ... Wait. I just got an email from Nancy.

> Hey Jer, I just read your "Ancient Rivalry." The answer is no. You can't.

Solution on p. 190

Hints for <u>Switch Central</u> (p. 62)

1. Can any Milwaukee player score twice in the eighth inning?
2. Who scored for Milwaukee in each inning?
3. In what order were the runs scored in Milwaukee's eighth inning?
4. Who pitched to each batter of Milwaukee's eighth?
5. Can you say who drove in each run of the eighth?
6. When did Hegan hit his home run?

7. What additional detail can you add to the story of Milwaukee's eighth?

Solution on p. 195

Hints for Leary's Bad Day (p. 71)

Jerry: It helps to note who can possibly bat in each of the first 5 innings (using that the Angels left only 3 on base).

A key fact is that 3 of the 9 runs given up by Leary were unearned. How could that have happened?

Solution on p. 201

Hints for Double Plays (p. 74)

Jerry: I need help. You've created a pretty deep problem. The usual methods don't seem to help.

Jim: What happens after Foster drives in a run?

Solution on p. 206

Hints for Mistakes Were Made (p. 75)

Jim: I know I have to look at the innings where runs were scored. But then what?

Jerry: Unearned runs have to be explained by errors or passed balls. There aren't any passed balls in this game.

If a batter hits a homer, is it possible that his own run is unearned? Is this possible if he is one of the first 3 batters in an inning? If so, how?

And what were the outs in the ninth inning?

Solution on p. 206

Hints for At the Dawn of Time (p. 85)

Jerry: Focus on Curry's 2 runs. What rule makes that possible?

Jim: It's possible to nail down who leads off each inning. That will help place some of the runs.

Solution on p. 214

Hints for Twins Fall Short in Slugfest (p. 92)

Jim: Okay, it's a trek. Start me off!

Jerry: Well, start with Mincher's homer. When was it and who was the pitcher then?

And for tracking runs, it's helpful to use the fact that the box score lists homers in the order that they were hit.

Jim: Do I need to use that?

Jerry: Ask Nancy.

Solution on p. 223

Hints for <u>In Scoring Position</u> (p. 94)

Start with the Twins' pitchers in the seventh and eighth innings—who did Cressend pitch to in the eighth? Who did Miller, Wells, and Mays pitch to in the seventh?

Solution on p. 225

Hints for <u>Two Guys</u> (p. 98)

Jim: Can you say anything more about this puzzle?

Jerry: Under what circumstances can Dominguez drive in Hall?

        Don't forget the Gap Rule, p. 24.

Solution on p. 230

Hints for <u>Castro Homers in the First</u> (p. 100)

Jim: Jerry, I ...

Jerry: Wait. Nancy saw my solution and found a better one. She suggests making a chart showing when everyone bats. Then focus on the players who don't have homers. You don't know who they are yet. But whoever they are, you can easily prove that they have to drive each other in.

Jim: Whoa.

Solution on p. 235

Hints for <u>Leftovers</u> (p. 102)

	ab	r	h	bi
Alonso ph, 2b	2	1	2	0
Chen rf	1	1	1	0
Croteau ph, rf	3	0	0	0
Faatz p	1	0	0	0
Gil ph, lf	1	0	1	1
Hanlon cf	3	1	2	1
Jackson p	0	0	0	0
Kelly lf	3	0	3	2
Kimura p	2	0	0	0
Maeng 2b	2	0	1	1
Pasig 3b	4	2	2	0
Tabeau 1b	3	0	1	1
Yoshida c	4	1	0	0
Zhao ss	3	0	1	0

	ip	h	r	er	bb	so
Martin (W)	9	14	6	6	0	4

LOB—Eastham 9, Westwood 5, HR—Croteau, Namath, Okada, DP—Westwood 6, 3B—Hanlon (2), HBP—Duffy by Faatz, GIDP—Pasig, Yoshida (3), Hanlon, Faatz, SF—Hanlon

Westwood   4 0 0 0 0 0 0 0 2—6

Jerry: I haven't got this figured out yet. I did figure out who scored in the first and who in the ninth.

Jim: And did you do the same for the rbis?

Jerry: Yeah.

Jim: If you have any trouble, the number left on base is useful.

Jerry: Thanks!

What I'm working on now is all those guys who grounded into double plays. I think I'm supposed to use them to prove stuff. That's kind of cheap, isn't it?

But in the back of my mind is the fact that if there are two men on base, a guy can hit into a double play and wind up safe on first. And then he could later score!

Jim: Oh.

Jerry: But great puzzle, Jim!

Solution on p. 238

Hints for Who Won? (p. 114)

Jerry: I got this far: Iglesias is listed as a batter for the Cats and is also listed first as a pitcher in the pitching section of the box score. Since visiting team pitchers are always listed first in the pitching statistics, we know that Cats must be the visiting team, and therefore bats in the first half of every inning. So Ito must the starting pitcher for the home team, Dogs.

Then, as I said on page 49, now what?

Jim: So here are a few questions to lead you on:

How many plate appearances did Iglesias have in innings 1–5?

Which batting slot leads off the sixth inning?

When did Ivory drive in a run?

Good luck!

Solution on p. 243

Hints for Kids' Stuff (p. 117)

Jim: Jerry, I need just a little something to get me started.

Jerry: Note the exact wording of the puzzle: "every batter is either out on strikes or reaches first on a walk." That means every batter who strikes out is actually out—he doesn't reach first on a dropped third strike. They don't allow that in most Little League games, anyway. Since walks don't count as at-bats, all 18 at-bats in the box score must be strikeouts, and this accounts for all 18 outs in 6 innings. So once a player walks, he can't be picked off base, or out stealing. He either scores or is left on base.

Jim: I see that every run was driven in. So that means that every run must have been scored on a bases loaded walk. A stolen base doesn't come with an rbi.

Jerry: Right. And since a run is scored in every inning on a walk, the bases must get loaded every inning and when the inning ends with a strikeout, the bases are still loaded. Thus there are 3 LOB in each inning in which a run is scored. Given the number of runs each inning, that tells us that every inning had 7 plate appearances (1 run, 3 outs, 3 left on base).

Jim: Stop! That's all I need!

Jim needed another hint. It's on p. 177.
The solution is still on p. 246.

Hints for Kids Will Be Kids (p. 118)

Jerry: Jim, as in the previous chapter, I suggest you work on filling out a scorecard. But notice that in the third inning, the Puppies scored 4 runs. With 3 LOB, that means 10 at-bats in the inning. That makes a single column for the third inning a little crowded. It's even worse for the second inning (6 runs, 12 at-bats). So to give you plenty of room, there are three columns for the second inning and two for the third.

Puppies	1	2	3	4	5	6	ab	R	H	rbi
Aiko	?	? K	?		?	?	1	4	0	1
Butch	?	?	? K		?	?	2	2	0	1
Chulo	?	?	?	?	?	?	3	3	0	0
Dusty	?	?	?	?	?	?	2	0	0	0
Em	?	?	?	?	?	?	1	1	0	4
Fei Fei	?	?	?	?	?	? K	4	2	0	1
Gordito	K	?	?	?	?		1	1	0	4
Hank		? ?	?		? K		2	1	0	2
Izzie		? ?	?	K		?	2	1	0	2
TOTALS	R/H 1/0	6/0	4/0	1/0	2/0	1/0	18	15	0	15

Solution on p. 248

More hints for Kids' Stuff (p. 117)

Jim: Jerry, here's what I did. I made up a scorecard for the game ("K" for strikeouts).

Puppies	1	2	3	4	5	6	ab	R	H	rbi
Aiko	?	?	?	K		?	2	1	0	0
Butch	?	?	?		?	?	2	0	0	2
Chulo	?	?	K		?	?	2	1	0	1
Dusty	?	?		?	?	?	2	1	0	0
Em	?	K		?	?	?	2	0	0	1
Fei Fei	?		?	?	?	K	2	1	0	1
Gordito	K		?	?	?		2	0	0	1
Hank		?	?	?	K		2	1	0	0
Izzie		?	?	?		?	2	1	0	0
TOTALS	R/H 1/0	1/0	1/0	1/0	1/0	1/0	18	6	0	6

I thought I'd fill in strikeouts for each player who ends an inning. Then I got stuck.

Jerry: A scorecard is good. Just keep in mind that a player who scores the first run in an inning must be one of the first three batting in the inning.

And a player can't get an rbi in an inning in this kind of game unless the bases are loaded. So a player can get an rbi in an inning only if he's batting fourth, fifth, or sixth in the inning.

Also, in order to have 2 at-bats in the game, a batter must strike out exactly twice and walk in all the remaining plate appearances.

Jim: Thanks!

Solution on p. 246

Hints for Zhang and Li (p. 121)

Jerry: For each team, could it have played a type A game?
        For each team, could it have played a type B game?
        For each team, could it have played a type C game?
        (see p.250)

You will need to look for 1-run games among 6-inning games, extra-inning games, walk-off wins, and rain-shortened games. Keep going. Don't give up. It's fun.

Solution on p. 251

Hints for Advanced ERGonomics (p. 123)

Jerry: There are three types of ERG games (p. 250):

Type A, where all nine players bat in each inning, type B, where the lineup splits in innings 1 and 2, 3 and 4, 5 and 6 in one of four ways: 3 in 1 inning and 6 in the next, 4 in 1 inning and 5 in the next, 5 and then 4, or 6 and then 3, and type C, where 3 batters bat each inning.

Could a game between a type A team and a type B team have a difference of 8 in PAs? What about a game between a two type B teams? Eliminate from consideration all match-ups that you can.

Solution on p. 255

Solutions

Solution to <u>Casey at the Bat</u> (p. 24)

Casey bats third. He's up four times. And he's the last batter. So each slot has at least 3 plate appearances and then the first 3 batters have one more plate appearance each for a total of $3 \times 9 + 3 = 30$ PA. PA is the sum of Runs, Outs, and LOB. We know Outs, that's 27. We know PA. If we knew LOB, we could calculate the number of runs.

In the ninth, Casey stranded three players—he struck out with the bases loaded, so LOB is at least 3. But LOB can't be bigger than 3 or we'll have Runs plus Outs plus LOB more than 30. So LOB is exactly 3. So Runs has to be 0. And since we're told the Slugs were 2 runs behind, they lost 2–0.

Solution to <u>Number 400</u> (p. 48)

In the game, Mantle scored twice and Richardson scored once. New York scored twice in the ninth inning. Mantle couldn't have scored twice in the inning; it had to be Richardson and Mantle scoring in the ninth. Mantle had only one rbi so he hit a solo home run. Could that have occurred in the ninth inning? Note that Richardson bats 2 slots before Mantle in the batting order. If Mantle homers in the

NEW YORK	ab	r	h	bi	DETROIT	ab	r	h	bi
Kubek,ss	4	0	0	0	Fernandez	4	0	1	0
Richardson,2b	4	1	2	0	Bruton,cf	4	0	1	0
Tresh,lf	3	0	0	0	Kaline,rf	3	1	2	1
Mantle,cf	3	2	2	1	Colavito,lf	4	0	1	0
Lopez,rf	4	0	1	1	cWood	0	0	0	0
aLinz	0	0	0	0	Cash,1b	3	0	0	0
Maris,rf	0	0	0	0	bMorton	1	0	0	0
Howard,c	4	0	2	1	McAuliffe,2b	3	0	1	0
Skowron,1b	4	0	0	0	Kostro,3b	3	0	0	0
Boyer,3b	3	0	0	0	Brown,c	3	0	0	0
Terry,p	3	0	0	0	Aguirre,p	3	0	0	0
Daley,p	0	0	0	0	Fox,p	0	0	0	0
Total	32	3	7	3	Total	31	1	6	1

aRan for Lopez in the 9th; bHit into force play for Cash in 9th; cRan for Colavito in 9th.

New York .... 0 0 0 0 1 0 0 0 2—3
Detroit .... 1 0 0 0 0 0 0 0 0—1

E—None 2. A—New York 14, Detroit 11. DP—Boyer, Richardson, and Skowron; McAuliffe, Fernandez, and Cash. LOB—New York 4, Detroit 4. 2B—Richardson, Howard. HR—Mantle, Kaline. SB—Fernandez. S—Tresh.

	IP.	H.	R.	ER.	BB.	SO.
Terry (W, 21—10)	8.2	6	1	1	1	5
Daley	0.1	0	0	0	0	0
Aguirre (L, 14—7)	8.1	7	3	3	1	8
Fox	0.2	0	0	0	0	0

Umpires—Smith, Rice, Paparella, Scar. Time—2.05. Attendance—22,810.

ninth, Richardson must score before he comes up, since Mantle has only 1 rbi. But the only players who could have driven in Richardson are Lopez and Howard, who bat after Mantle clears the bases, so that is not possible. So Mantle's home run came in the fifth inning.

Consider the first 4 innings. By the PA equation for those innings, PA = 12 outs + 0 runs + 0–4 LOB = 12–16. That means that the player leading off the fifth inning batted in slot 4, 5, 6, or 7. Mantle batted fourth, so the only way he could have homered in the fifth inning was if he led off the inning.

Note: Usually home runs are listed in the box score in the order in which they are hit. In this case, though, Mantle's fifth-inning homer is listed before Kaline's first-inning homer. That's probably a mistake.

Solution to <u>Who Scored When</u> (p. 49)

The Mets have 5 runs, 6 left on base, and 24 outs (from 8 full innings). By the PA equation, that's 35 plate appearances. Every slot of the nine in the lineup should have 4 PA except the last slot, which should have 3 ($35 = 8 \times 4 + 3$). Note, however, that there were only 29 at-bats and that slots 2, 3, 5, and 6 appear to be missing a PA and that slot 8 is missing two PA.[1] But DeLeon gave up 4 walks, DeLeon hit Samuel (HBP), and Teufel had a sacrifice fly (SF). Each of those are plate appearances that aren't counted as at-bats. Thus, Johnson and McReynolds had walks and Elster had two.

With 35 PA, the final plate appearance is Elster's in slot 8. Before him is Carter, who has the last plate appearance of slot 7. Carter has only one PA, so his rbi single (he has no 2B, 3B, or HR) was in that inning, the eighth.

The previous batter in the eighth, Teufel, also has the final PA of his slot. He too has only one PA. That must be his sacrifice fly which must drive in the second run of the eighth. Whom did he drive in? Magadan? McReynolds? We'll see in a moment.

McReynolds is special. He has a home run but just one rbi, so it must be a solo homer. And McReynolds scores twice. He must score in both the second and eighth innings.

---

[1] Slot 6 is occupied first by Magadan and then later by Teufel, slot 7 is occupied first by Sasser, then Carter.

McReynolds can't score on his second PA, the 14th of the game. With one more out needed, that would require 15 batters in the first two innings: 6 outs, 3 runs, and all 6 LOB. But when he scores, nobody is left on base, so that's impossible. Thus he scores on his first PA, and can't bat in the first inning. The first inning must consist of Miller, Samuel, Johnson, and possibly Strawberry. McReynolds scores the first run of the second inning.

The other 2 runs of the second inning must be scored by Magadan and Elster. Johnson can't score in the second because all the non-scorers preceding him would have to be out and that would include Sasser, Darling, Miller, and Samuel (more than 2—impossible). So the runs in the second are scored by McReynolds, Magadan, and Elster. That leaves Johnson and McReynolds to score in the eighth.

Who drives in whom and when does McReynolds homer? Every run is driven in (there are 5 runs and 5 rbis). So who drives in Johnson in the eighth? Not McReynolds, whose sole rbi was on his homer, driving in himself. It must be that Teufel drives in Johnson (while McReynolds is on base) with a sacrifice fly. Then Carter drives in McReynolds.

In the second inning, McReynold drives in himself. That leaves Miller to drive in Magadan and Elster. Is that possible? It is. Suppose that McReynolds leads off the second inning. Then Magadan gets on base. Then Sasser makes an out. Then Elster gets on, perhaps a walk.

FIRST GAME								
ST. LOUIS				**METS**				
	ab	r	h	bi		ab	r	h	bi
Coleman, lf	5	0	0	0	Miller, 2b	4	0	1	2
O. Smith, ss	4	0	1	1	Samuel, cf	3	0	1	0
M. Thmp, cf	3	0	0	0	H. Jhnsn, 3b	3	1	1	0
Guerrer, 1b	4	0	2	0	Strwbry, rf	4	0	0	0
Pndltn, 3b	4	0	1	0	McRylds, lf	3	2	1	1
Oquend, 2b	3	0	1	0	Magadn, 1b	3	1	1	0
Brnnsky, rf	4	0	1	0	Teufel, 1b	0	0	0	1
Dayley, p	0	0	0	0	Sasser, c	3	0	0	0
Quisnbry p	0	0	0	0	Carter, c	1	0	1	1
T. Pena, c	2	1	0	0	Elster, ss	2	1	0	0
Walling, ph	1	0	0	0	Darling, p	3	0	1	0
Pagnozzic, c	1	0	0	0	Totals	29	5	7	5
DeLeon, p	2	0	1	0					
Morris, rf	1	0	0	0					
Totals	34	1	7	1					

St. Louis .... 0 0 1 0 0 0 0 0 0—1
Mets .... 0 3 0 0 0 0 0 2 x—5

E—Miller. LOB—St. Louis 9, Mets 6. 2b—Miller, O.Smith, H.Johnson. HR—McReynolds (13). SB—Oquendo (2), McReynolds (13). S—DeLeon. SF—Teufel.

St. Louis	IP.	H.	R.	ER.	BB.	SO.
DeLeon L. 11-11...	7.1	6	5	5	4	8
Dayley.............	0.1	0	0	0	0	0
Quisnbry..........	0.1	1	0	0	0	0
Mets						
Darling W. 10-9....	9	7	1	0	1	3

HBP—Samuel by DeLeon, M. Thompson by Darling. BK—Darling. Umpires—Home, Marsh; First, Hohn; Second, Wendelstedt; Third, Darling. T—2:49.

Then Darling either gets on or is retired. Then Miller drives in Magadan and Elster.

Solution to Earning and Unearning (p. 50)

For Los Angeles, the usual PA = R + outs + LOB equation gives 5 + 26 + 5 = 36 PA, so every slot in the batting order has 4 PA, and the last batter is Sudakis, the final player in the 9 slot. The 26 outs come from the 8 and 2/3 innings pitched by Decker and Regan.

The PA = AB + HBP + SF + S + BB equation, for each slot in the lineup, gives:
Wills must have a BB to make 4 PA. Mota' SF + 3 AB accounts for his 4 PA. Davis' S + 3 AB accounts for his 4 PA. The next 4 slots have 4 AB, and need no more explanation.

Joshua was a pinch runner for Haller after his fourth PA, so he must have pinch-run in the ninth and scored then.

Lefebvre started in slot 8. Since Singer was the starting pitcher, he couldn't have substituted for Lefebvre. Therefore, Lefebvre must have a BB to account for the 4 PA in slot 8.

This leaves 6 players in the 9 slot, from Singer to Sudakis, with only 3 AB and no HBP, no S, and no SF. One of them had a walk. It must be Peña, because he pitched 5 innings, so he must have had a PA, but he had no AB to account for that PA in any other way. That means that the 4 PA in the 9 slot were one each by Gabrielson, Peña, Crawford, and Sudakis.

Now that this is sorted out, it is easy to identify when each player's runs and rbis took place.

Los Angeles had a walk-off win in the ninth. Since Sudakis batted just once, he hit his homer then, and it was a 2-run walk-off homer, driving in Joshua, who we already proved scored in the ninth. Joshua pinch-ran for Haller, batting seventh, who could have reached base either on a hit or on a fielder's choice.

Since the first run in any inning must be scored by one of the first 3 batters, Wills must have scored the sole run in the first inning. He can't have been driven in by Haller, batting seventh, because they are too far apart. Mota must have driven in Wills and since he had just 1 rbi, it must have been from his sacrifice fly.

There are only 2 more Dodger runs to account for, by Kosco and Haller, so they must have scored in the second, and only 2 more rbis, to account for, by Haller and Gabrielson. Since the first 4 batters had PA in the first inning, Kosco must have led off the second inning. Haller, batting ahead of Gabrielson, must have driven in Kosco, and Gabrielson must have driven in Haller. Since Gabrielson had no hit, sacrifice, or sacrifice fly, he must have driven in Haller with a ground ball on which he did not get a hit. It might have been a ground ball out, or he might have reached first safely on a fielder's choice.

The Cubs' runs are easily identified, too. Popovic and Hickman are the only batters who could have scored in the first inning, and that leaves Banks and Hiatt to score in the second inning. Banks had a homer, so he drove himself in with a homer in the second inning, and then Hiatt reached base on a walk. Popovich didn't have a homer, so he couldn't have had an rbi in the first inning. His rbi must have

CHICAGO	ab	r.	h.	bi
Kessinger, ss	4	0	0	0
Popovich, 2b	4	1	3	1
Williams, lf	4	0	0	0
Hickman, rf	5	1	2	2
James, cf	0	0	0	0
Calison, rf	4	0	1	0
Sento, 3b	3	0	0	0
Banks, 1b	4	1	1	1
Hiatt, c	3	1	0	0
Decker, p	1	0	0	0
Regan, p	1	0	0	0
Total	33	4	7	4

LOS ANGELES	ab	r.	h.	bi
Wills, ss	3	1	0	0
Mota, lf	3	0	1	1
Davis, cf	3	0	0	0
Parker, 1b	4	0	0	0
Kosco, rf	4	1	1	0
Gr'b'k'witz, 3b	4	0	1	0
Haller, c	4	1	1	1
Joshua, pr	0	1	0	0
Lefebvre, 2b	3	0	1	0
Singer, p	0	0	0	0
Gabrielson, ph	1	0	0	1
Pena, p	0	0	0	0
Crawford, ph	1	0	1	0
Brewer, p	0	0	0	0
Sudakis, ph	1	1	1	2
Total	31	5	7	5

Chicago .... 2 2 0 0 0 0 0 0 0—4
Los Angeles .... 1 2 0 0 0 0 0 0 2—5

E—Decker. LOB—Chicago 10, Los Angeles 5. 2B—Kosco. HR—Hickman (16), Banks (6), Sudakis (4). SB—James. S—Decker, Davis. SF—Mota.

	IP.	H.	R.	ER.	BB.	SO.
Decker	6.2	4	3	2	3	3
Regan L (4-2)	*2	3	2	2	0	0
Singer	2	3	4	4	2	1
Pena	5	3	0	0	3	1
Brewer W (2-1)	2	1	0	0	1	3

*Two out when winning run was scored.
HBP—by Singer (Popovich)
Time—2.34. Attendance—28,756.

been in the second inning driving in Hiatt. So the 2 rbis in the first inning must have been by Hickman, and that's when he must have hit his homer, a 2-run homer.

Finally, how could it be that a different run was unearned and Wills' run earned? For a later unearned run, it's possible that in the second inning Decker made an errant throw that should have picked off Haller, causing Haller's run to be unearned.

One way Wills' run might be earned is that Mota's SF drives in Wills from first base when the outfielder is stunned or injured catching the ball. Another way is that Wills could advance as follows: a pitch by Decker is bobbled by the catcher, inducing Wills to start for second. But Hiatt quickly fires the ball to second base, and Wills is caught in a rundown. They could have chased him back to first, but instead he eludes the rundown and reaches second base. There is no passed ball, because Wills didn't reach second because of the bobble. There is no stolen base, because he didn't take off with the pitch. There is no trace of this event in the box score. But still he gets to second base. He could get to third the same way!

Solution to A Steady Game (p. 55)

Since Team B won, but didn't bat in the ninth inning, it must have scored at least one run in one of the first 8 innings. In that inning they must have had at least 4 PA and Team A must have had at least 3 PA, for a total of at least 7 PA. So the total PA in the ninth inning must also be at least 7, and they were all with Team A. With 7 PA, Team A must have scored at least one run.

Solution to Another Steady Game (p. 56)

Recall that the Bombshells played a steady game, scored 2 runs, and left 7 runners on base. They sent the same number of batters to the plate each inning. It must be 4. If it was 3, they couldn't have any men left on base and if it was 5, they would either have many more left on base or more than 2 runs.

The Bombshells had 4 PA in each inning and batted in 8 or 9 innings (either 32 plate appearances in 8 innings or 36 PA in 9 innings). Eight innings, however, is impossible since there would be 24 outs, 2 runs, and 7 LOB for $24 + 2 + 7 = 33$ PA. Instead, if they play in the ninth inning they can get to 36 PA—but only if they have 27 outs ($27 + 2 + 7 = 36$). In other words, they must have had 3 outs in the ninth. That means they lost the game; if they had won, the winning run would have been scored with fewer than 3 outs.

Solution to A Steady Double (p. 57)

Let R be the number of runs Brooklyn scores in each game (recall the score of both games is the same). Let p be the number of players Brooklyn sent to the plate each inning (again, the same in both games). Note that as visitor, Brooklyn had 27 outs in the first game and 30 in the second. Now in the first game we have:

$$
\begin{aligned}
\text{PA} &= R + \text{LOB} + \text{Outs} \\
9p &= R + 10 + 27 = R + 37
\end{aligned}
$$

and in the second, we have

$$
\begin{aligned}
\text{PA} &= R + \text{LOB} + \text{Outs} \\
10p &= R + 13 + 30 = R + 43
\end{aligned}
$$

Subtracting the 9p equation from the 10p equation gives us

$$ p = 6. $$

Both equations now tell us that R is 17 runs.

Solution to Cicioneddos (p. 57)

Playing a steady game, Brooklyn had a constant p plate appearances in each inning they batted. As the visiting team they made 3 outs in each of these innings. We have, per inning:

$$ p = 3 \text{ (Outs)} + \text{Runs} + \text{LOB, so:} \quad p - 3 = \text{Runs} + \text{LOB}, $$

a constant sum. Runs + LOB per inning must be 1, and 6 innings are needed to accumulate 2 runs and 4 LOB. (If Runs + LOB per inning were 2, a minimum 4 innings would result in too many runs or LOB.)

Solution to Maltagliatti (p. 58)

Let n stand for the number of innings played and p be the number of players Brooklyn sent to the plate in each inning. We have first that $\text{PA} = pn$. Sellary, in slot 9 had at least 4 at-bats so PA is at least 36.

Could *p* be 6? *n* would have to be at least 6. But in every inning where Brooklyn doesn't score (at least 5 innings), it leaves 3 on base. That's 15 LOB. Not possible.

Could *p* be 5? *n* would have to be at least 8. In every inning where Brooklyn doesn't score (at least 7), it leaves 2 on base. That's 14 LOB. Not possible.

p can't be 3 (it couldn't achieve 13 LOB). The only possibility left is 4.

With $p = 4$, the number of plate appearances for the Bombshells is $4n$, where *n* is the number of innings. This will have to be an extra-inning game, since in any complete inning the Bombshells can leave at most one on base.

Can Brooklyn win in extra innings? The home team only wins an extra-inning game in a walk-off. In this case, since Sellary drives in the Bombshell's only run, she has to be the last Bombshell at-bat. She's ninth in the batting order, so the number of plate appearances must be a multiple of 9. But when is $4n$ a multiple of 9? Only when *n* is a multiple of 9, that is, 9 innings (not enough to accumulate 13 LOB) or 18 innings (too many!) or more. That's all impossible. So Brooklyn loses the game. They lost in 14 innings—in one, Mavis drove in a run, in the other 13 the Bombshells left a player on base.

Solution to Lucca (p. 58); (Hint on p. 169)

Consider the losing team. We are told it has no runs. Since it lost, it had a full 3 outs in every inning. To have the same number of batters every inning (and no runs) it must have the same number of LOB, either 0, 1, 2, or 3, in each inning. If it's Brooklyn, the only way it could reach the right number of LOB would be to have 1 LOB for 11 innings. If it's Lucca, the only way would be to have 1 LOB for each of 15 innings or 3 LOB for each of 5 innings (but Lucca can't have 5 LOB for each of 3 innings because there are only 3 bases.).

The case of Brooklyn losing in 11 innings fails, because the score would be 0-0 after 10 innings (or the game wouldn't get to the eleventh inning). Lucca would have to have a constant number of LOB in those 10 innings. It couldn't be 2 or more per inning, or

Lucca would have 20 or more LOB. It couldn't be 1 per inning, or Lucca would have 10 LOB after 10 innings and would need an impossible 5 more in the 11th. It certainly couldn't be 0 LOB for 10 innings, leaving 15 LOB for the 11th inning!

The case of Lucca losing in 15 innings also fails, because the score would be 0-0 after 14 innings (or else the game wouldn't go to the 15th inning). Brooklyn would have to have a constant number of LOB in those 14 innings. It couldn't be 1 or more or Brooklyn would have more than 11 LOB. And it couldn't be 0—that would leave 11 LOB for the 15th inning.

The only possibility left is that Lucca loses in 5 innings, a rain-shortened game. Can we arrange for Brooklyn to have its 11 LOB and 58 runs in such a game?

Either Brooklyn is the visiting team or the home team. We'll try visiting team first.

Say Brooklyn had s PA per inning. Then Brooklyn had a total of $5s$ PA, equaling runs+outs+LOB. Then we have

$$PA = 5s = 58 + 15 + 11 = 84.$$

That's simply not possible since 84 is not a multiple of 5.

Now we'll try Brooklyn as the home team. There are two possibilities: The game ends after 4 1/2 innings or the game ends with Brooklyn batting in the fifth inning. In the first case, we have the equation

$$PA = 4s = 58 + 12 + 11 = 81.$$

That's not possible; 81 is not a multiple of 4.

In the second case we have the game ending in the fifth with Brooklyn having 0, 1, 2, or even 3 outs. The equation would be

$$PA = 5s = 58 + \text{Outs} + 11,$$

where Outs is 12, 13, 14, or 15. The sums in those cases are: 81, 82, 83, 84. None of these numbers is a multiple of 5. Our conclusion is *No*, the winning team cannot score 58 runs.

Extra: The winner can't score 58, but it turns out that 57 and 59 are both possible: Brooklyn could score 59 runs by playing 5 innings as visitor, because $59 + 15 + 11 = 85$ with Brooklyn having 17 plate appearances per inning. Note that Lucca would play 5 innings bringing 6 players to the plate each inning with 3 of them left on base each inning for a total of 15 LOB.

Finally, Brooklyn could play 4 innings (as the home team, with Lucca playing 5 innings) scoring 57 runs because $57 + 12 + 11 = 80$ with Brooklyn having 20 plate appearances per inning.

Solution to Ancient Rivalry (p. 59); (Hint on p. 170)

First note that Boston has the same number of runs and rbis, so each run was driven in by someone, and did not score in some other way, such as on an error. At the same time, five of its runs were unearned.

In what innings does Yastrzemski score?

Yastrzemski scores 3 runs. They must be in different innings (scoring twice in one inning requires 8 runs in the inning—Gap Rule, p.24). Thus, Yastrzemski scores in the first, the fourth and the fifth.

When does Bressoud drive in runs?

Bressoud drives in himself with a homer and someone else with a sacrifice fly. Since Yastrzemski scores the only run of the first before Bressoud comes to the plate, Bressoud doesn't do either (HR or SF) in that inning.

YANKEES

FIRST GAME

BOSTON	ab	r	h	bi	NEW YORK	ab	r	h	bi
Geiger,cf	4	1	0	0	Kubek,ss	3	2	1	0
Bressoud,ss	4	1	1	2	Richardson,2b	4	1	2	0
Yastrzemski,lf	4	3	2	2	Tresh,lf	3	0	2	1
Clinton,rf	5	1	3	2	Maris,cf	3	0	1	2
Runnels,1b	5	1	1	2	Lopez,rf	4	0	0	0
Malzone,3b	4	0	1	1	Howard,c	4	0	1	0
Nixon,c	4	0	1	0	Skowron,1b	4	0	0	0
Schilling,2b	4	1	2	0	Boyer,3b	4	0	0	0
Monb'quette,p	2	1	0	0	Brown,p	1	0	0	0
Total	36	9	11	9	Sheldon,p	0	0	0	0
					aLinz	1	0	0	0
					Cullen,p	0	0	0	0
					bBlanchard	1	0	0	0
					Clevenger,p	0	0	0	0
					cLong	1	0	0	0
					Total	33	3	7	3

aFlied out for Sheldon in the 5th; bFouled out for Cullen in the 7th; cFouled for Clevenger in the 9th.

Boston .... 1 0 0   3 5 0   0 0 0—9
New York .... 2 0 0   0 1 0   0 0 0—3

E—Boyer, Skowron. A—Boston 7, New York 11. DP—Malzone, Runnels; Zipfel; Clevenger, Kubek, Skowron. LOB—Boston 5, New York 6. 2B Hits—Yastrzemski, Richardson, Maris, Tresh. 3B—Malzone, Clinton. HR—Yastrzemski, Bressoud, Clinton. Sacrifices—Monbouquette, Geiger. SF—Bressoud.

	IP.	H.	R.	ER.	BB.	SO.
Monb'q'ette (W, 12—13)	9	7	3	3	3	6
Brown (L, 6—5)	4.2	7	9	4	1	1
Sheldon	0.1	0	0	0	0	0
Cullen	2	1	0	0	1	2
Clevenger	2	3	0	0	0	1

Umpires—Paparella, Soar, Smith, Rice. Time—2.37.

Suppose Bressoud does both (HR and SF) in the fifth. If he hits his HR second, then everybody before his homer must score except for 2 batters who can be out. That produces too many runs. Impossible.

Suppose instead he hits his sacrifice fly second. You can't get credit for a SF unless there are fewer than 2 outs when you hit it. Thus when Bressoud comes to the plate, there is at most 1 preceding out, 3 men on base, and 4 runs in. 1+3+4 is not enough to account for 9 preceding batters. This is also impossible.

Similarly, Bressoud can't have both his SF and HR in the fourth inning. So he must have his HR and SF in separate innings, one in the fourth, and one in the fifth.

Who leads off the fourth inning?

In the first 3 innings, there is 1 run, 9 outs, and 0-5 LOB for a total of 10–15 PA. That means leading off in the fourth is either Bressoud, Yastrzemski, Clinton, Runnels, Malzone or Nixon. Yastrzemski must bat in the fourth inning and score. It can't be on his third time up, the 21st PA of the game. He'd be at least the 6th batter, requiring at least 4 runs in the inning. So Bressoud or Yaz must lead off. Bressoud has an rbi in the inning. If Yaz leads off, 8 batters will precede Bressoud. Two can be out, and 3 can be on base, so 3 will have already scored. Bressoud's rbi would be a 4th run. So Bressoud must lead off.

When do Bressoud and Yastrzemski hit homers?

Bressoud can't hit a sacrifice fly leading off in the fourth, so he does that in the fifth and so he homers in the fourth. Since Yastrzemski's homer must precede Bressoud's, Yastrzemski's homer is in the first.

When did Schilling score his run?

If Schilling scores in the fourth inning, then since he is the seventh batter of the inning, at least 5 (Gap Rule) of the preceding batters must score, whereas there are only 3 runs in the inning. So he can't score in the fourth and must score in the fifth. The same logic shows that Monbouquette and Geiger can't score in the fourth, and also must score in the fifth.

Who drove in whom in the fifth inning?

In the fifth inning, Schilling, Monbouquette, Geiger, and Yasztremski must all score. Furthermore, they have to score in that order. Any other order has three intermediate non-scorers, when the maximum possible is two. In addition, the only way for Schilling to bat early in the inning is if the fourth inning has the minimum number of batters (no LOB, 3 runs, 3 outs) starting with Bressoud and ending with Nixon. So Schilling must lead off the fifth inning.

The first player with an rbi is Bressoud, so Schilling, Monbouquette, and Geiger must load the bases for Bressoud to drive in Schilling with his SF. Yastrzemski now must get on base. At this point Yaz has had one rbi in the first inning from his homer, and no rbis in the fourth because he batted with the bases empty after Bressoud's homer, so he must drive in Monbouquette for his second rbi.

Clinton now comes to the plate and Geiger and Yastrzemski must still be on base in order to score later. Clinton can't have his homer now, because it would be a 3-run homer when he only had 2 rbis in the entire game. So Clinton must have his homer in the fourth inning instead. It must be a 2-run homer in the fourth, because we know Yaz was driven in by someone in the fourth and he batted just before Clinton's homer. So Clinton had no rbi in the fifth, and the only remaining possibility is that Runnels drove in Geiger and Yaz, followed by Malzone driving in Runnels.

The following scorecard shows everything we proved plus a few strong possibilities (Malzone's triple, Yastrzemski's double, and the errors making all the fifth-inning runs unearned). But remember, the point of making a scorecard here is to show that our analysis is logically consistent, that there really does exist a game that satisfies everything in the box score plus everything we have deduced. We're not claiming that *everything* we put in this scorecard actually happened that way. So, for example, one of the errors in the fifth inning could have been on Schilling's plate appearance rather than, say, Geiger's.

Boston	1	2	3	4	5	6	7	8	9	10	11	ab	R	H	rbi
Geiger	•		•		e							4	1	0	0
Bressoud	•			HR	SF							4	1	1	2
Yastrzemski	HR			2								4	3	2	2
Clinton	•			HR	•							5	1	3	2
Runnels		•		1								5	1	1	2
Malzone		•		3								4	0	1	1
Nixon		•		•								4	0	1	0
Schilling			•	1								4	1	2	0
Monb'quette			•	e								2	1	0	0
TOTALS	R/H 1			3	5							36	9	11	9

Jerry: Nancy points out that it would be impossible to solve the problem without using the order of homers to conclude that Yaztremski homered in the first. Without this restriction, the scorecard below shows a different scenario in which Yaz doubled in the first, after which Clinton drove him in with a single and was then thrown out for the third out.

Boston	1	2	3	4	5	6	7	8	9	10	11	ab	R	H	rbi
Geiger	•		•		⟨e⟩							4	1	0	0
Bressoud	•			HR	SF							4	1	1	2
Yastrzemski	⟨2⟩			HR								4	3	2	2
Clinton	1			HR	•							5	1	3	2
Runnels		•		•	⟨1⟩							5	1	1	2
Malzone		•		•	3							4	0	1	1
Nixon		•		•								4	0	1	0
Schilling			•		⟨1⟩							4	1	2	0
Monb'quette			•		⟨e⟩							2	1	0	0
TOTALS	R/H 1			3	5							36	9	11	9

Solution to <u>Double Switches</u> (p. 60)

<u>When did Floyd replace Johnson?</u>
Baltimore had 27 outs, 6 runs, 9 LOB for 42 PA. That means the first 6 slots had 5 plate appearances each; the 7–9 slots had 4. In slot 6, Johnson had 1, Floyd had 3 (he had a sacrifice) and May had 1. With Baltimore's 4 runs in the first inning, Baltimore had at least 16 PA in the first 4 innings. Slot 6's second plate appearance was Baltimore's 15th, so Floyd had to replace Johnson after the first and no later than the fourth inning.

<u>When did May pinch-hit for Floyd?</u>
May is the final batter in slot 6, so he has the last PA of the game for Baltimore. That's in the ninth inning.

<u>When the runs were scored:</u>
Baltimore scores 4 runs in the first inning. By the Gap Rule (p. 24), the 4 scorers must be in the first 6 slots. That means they have to be Buford, Salmon, Robinson, and Rettenmund. Buford, Robinson, and Rettenmund each have only one run, so these must be in the first inning. The 2 runs in the sixth must be Salmon's second run and Belanger's only run.

Rettenmund bats 6 slots after Belanger—too far away to drive him in. So Salmon must drive in Belanger in the sixth inning, and it must have been with his homer. Otherwise his homer would be in the first, driving in Buford. In that case Rettenberg could have only 2 rbis in the first inning, and at most 1 in the sixth, when we know he has 4 rbis. So Salmon's has 2 rbis in the sixth, and all 4 rbis by Rettenmund are in the first inning—a grand slam home run.

MILWAUKEE	ab	r.	h.	bi	BALTIMORE	ab	r.	h.	bi
Harper, 3b	5	2	2	0	Buford, lf	4	1	1	0
Hegan, 1b	4	3	2	2	Salman, 3b	5	2	2	2
Savage, cf	5	1	2	2	F.Robinson, rf	3	1	1	0
Walton, lf	3	0	0	0	Powell, 1b	5	0	0	0
Francona, 1b	1	1	1	1	Rettenm'nd, cf	4	1	2	4
Hershberg'r, rf	3	0	0	0	Johnson, 2b	1	0	0	0
Snyder, lf	1	0	0	0	Floyd, 2b	2	0	0	0
Kubiak, ss	3	1	1	0	May, ph	1	0	0	0
Roof, c	2	0	0	0	Hendricks, c	3	0	1	0
Humphreys, p	0	0	0	0	Richert, p	0	0	0	0
Pena, ss	2	1	1	3	Hall, p	0	0	0	0
Gil, 2b	1	0	0	0	Belanger, ss	4	1	2	0
Alvis, ph	1	0	0	0	Cuellar, p	3	0	1	0
Sanders, p	1	0	0	0	Watt, p	0	0	0	0
Baldwin, p	0	0	0	0	Dalrymple, c	1	0	0	0
Bolin, p	1	0	0	0	Total	36	6	10	6
McNertney, c	3	0	1	1					
Total	36	9	10	9					

Milwaukee .... 2 0 0   0 0 1   0 6 0—9
Baltimore .... 4 0 0   0 0 2   0 0 0—6

E—Johnson, Belanger. LOB—Milwaukee 3, Baltimore 9. 2B—Salman, Pena. HR—Rettenmund (9), Hegan (4), Salman (2). SB—Belanger, Savage, Francona. S—Floyd.

	IP.	H.	R.	ER.	BB.	SO.
Bolin	3.1	6	4	4	2	5
Humphreys	2.2	4	2	2	1	3
Sanders (W, 1—0)	1	0	0	0	0	0
Baldwin	2	0	0	0	1	1
Cuellar	7	7	6	5	2	7
Watt (L, 2—3)	0	1	1	1	0	0
Richert	0.2	2	2	2	1	2
Hall	1.1	0	0	0	0	1

HBP—by Sanders (Rettenmund).
T—2:55. A—6,280

Solution to <u>Switch Central</u> (p. 62); (Hint on p. 170)

Answering the questions in the Hint:

1. Can any Milwaukee player score twice in the eighth inning?
No. By the Gap Rule (p.24).
2. Who scored for Milwaukee in each inning?
Since nobody scored twice in the eighth, all 6 Milwaukee players
who scored, namely Harper, Hegan, Savage, Francona, Kubiak, and
Peña must have scored once each in the eighth. Hegan scored 3 runs,
and didn't score twice in any inning, so he must have scored in all 3
innings with runs, the first and sixth as well as the eighth. That leaves
Harper to score the other run in the first.
*3. In what order were the runs scored in Milwaukee's eighth
inning?*
Six players must score before the third out.

slots⟶	1	2	3	4	5	6	7	8 9
scorers	Harper	Hegan	Savage	Francona		Kubiak	Peña	

The possible lead-off slots that allow all six to score are 1, 6, or 9.
The last slot to bat in the game is 3. With 6 runs, 6 outs, and 0–3 LOB
in innings 8–9, the possible lead-off slots in the eighth are 7, 8, 9, and
1, so it's either 9 or 1. More on this later. In either case, Harper scores
first and Peña last. Peña is driven in by McNertney who also
(somehow) makes the final out in the eighth, leaving just 3 batters for
the ninth. More on this later.
4. Who pitched to each batter in Milwaukee's eighth?
Cuellar is charged with three of the runs in the eighth but none of
the outs. So he must have let the first three runners get safely on base.
Could the first runner have been McNertney? Maybe, as long as he
was put out when another pitcher was in the game. But no, Watt also
lets one runner on base and gets nobody out. So that would have
scored McNertney, who in actuality got no runs. So McNertney didn't
lead off the eighth: Harper led off. Cuellar pitched to Harper, Hegan,
and Savage.
Watt replaced Cuellar. He is charged with 1 run and no outs, so he
must have put the next scorer, Francona (Watkins' replacement), on

base. Watt's subsequent replacement, Richert, struck out two, walked one, and gave up two hits. That's 5 batters ending with McNertney. That's the rest of the batters in the eighth inning, so Watt pitched only to Francona.

Richert pitched to the next 5 batters. More on this later. That's all the batters in the eighth. But Richert only recorded 2 outs. The third out belongs to Hall, but Hall didn't pitch to a batter in the eighth! How is this possible? What must have happened is that after McNertney drove in Peña, Richert was removed for Hall who then faced Harper. But before Harper's at-bat was finished, McNertney was retired. Either McNertney was picked off base or was caught stealing. So Hall was credited with 1 out in the eighth, although he didn't (officially) deal with any at-bats in the eighth!

5. *Can you say who drove in each run of the eighth?*

We have already noted that McNertney drove in Peña.

Peña has 3 rbis and 2 plate appearances. Since Harper led off the eighth inning, Peña must have had his other plate appearance in the seventh inning. So he didn't bat in the first or sixth innings and didn't have any rbis then. Working backward from McNertney, Peña was the next to drive in runs, so he must have driven in the previous three scorers, Savage, Francona, and Kubiak.

There are only two left to drive in, Harper and Hegan. Francona only comes into the game in the seventh inning and has 1 rbi so he had to drive in one of them. It has to be Harper.

Since Milwaukee had an equal number of runs and rbis, every batter who scored was driven in by somebody. Hegan didn't drive himself in with his homer, because we just proved that Francona drove him in. That leaves only Savage who could have driven in Hegan, so he did.

Harper could have been driven in by Hegan or Savage. That's also true in the first inning. It doesn't seem possible to decide which.

6. *When did Hegan hit his home run?*

We already showed he didn't hit his homer in the eighth. Hegan's homer comes after Rettenmund's, so it couldn't have occurred when Milwaukee, as visiting team, had their turn to bat in the first inning before Baltimore. Thus, Hegan homered in the sixth inning.

7. *What additional detail can you add to the story of Milwaukee's eighth?*

Savage and Francona stole bases. Francona's steal must have been in the eighth inning, the only time he was on base. Harper and Hegan had scored, but not Savage, so Francona was on first. Either Savage was on third or else this was a double steal. Both situations are possible.

In addition to striking out Snyder and Sanders, we can deduce that Rickert walked Kubiak and yielded a double to Peña and a single to McNertney.

For the sixth, Roof had 2 at-bats before being replaced as catcher after the fourth inning; hence the PA for innings 1–4 was at least 16 and thus the PA for innings 1-5 was at least 19. But it was at most 19 for Hegan to homer; hence it was exactly 19 and Hegan led off the sixth.

Solution to A World Series Game (p. 65)

1. By the Gap Rule, the only player who can score in the first is Oberkfell, so he does. We now have two possible cases: (a) Oberkfell scores his other run in the sixth, in which case both Smiths and McGee score in the second, or (b) Oberkfell scores in the second. We will rule out case (a).

2. Assuming case (a), it's L.Smith, McGee, and O.Smith who score in the second inning.

3. The first 5 innings have 4 runs, 15 outs, and 0–6 LOB, for 19–25 PA. So anybody from the 20th batter, Oberkfell, in slot 2, through McGee, in slot 8, could lead off the sixth.

4. But by the Gap Rule, Oberkfell can't score in the sixth if McGee leads off. In fact, he can only score if he's the one leading off the sixth.

5. But Oberkfell leading off the sixth inning means that LOB is 0 for the first 5 innings. So in the first inning there are only 4 batters. So Porter leads off the second inning.

6. There are 5 batters before Herr drives in somebody in the second: Porter, Iorg, L.Smith, McGee, and O.Smith. At most three can be on base. At most one can be out (there can't be more than one out when Herr hits his sacrifice fly). So one run is in. That has to be L.Smith. So setting the stage, Porter is out. L.Smith is in. And the bases are loaded with Iorg, McGee, and O.Smith.

7. Now Herr hits a fly ball. The runners can't run until the ball is caught. That's the second out. But McGee and O.Smith can't score until Iorg is out. That's the third out. Impossible. Case (a) leads to a contradiction.

8. Thus, Oberkfell can't score in the sixth inning and must score in the second inning.

9. The 3 runs in the second are Oberkfell plus 2 runs driven in by Herr's sac fly. That means that Iorg, who has an rbi, must drive in his run in the sixth. By the Five-step Rule, he can only drive in Oberkfell or L. Smith. Since Oberkfell doesn't score in the sixth, Iorg drives in L. Smith in the sixth.

10. We have, finally, that Herr drives in McGee and O.Smith with his sacrifice fly.

11. Oberkfell scores in the first and the second innings. In one of those he is driven in by Hendrick. In the other, he scores without being driven in. But we can't tell which!

Brewers 7, Cardinals 5

GAME FOUR

ST LOUIS	ab	r.	h.	bi	MILWAUKEE	ab	r.	h.	bi
Herr 2b	4	0	0	2	Molitor 3b	4	1	0	0
Oberkfell 3b	2	2	1	0	Yount ss	4	1	2	2
Tenace ph	1	0	0	0	Cooper 1b	4	1	2	1
Hrnandz 1b	4	0	0	0	Simmons c	2	0	0	0
Hendrick rf	4	0	1	1	Thomas cf	4	0	1	2
Porter c	3	0	1	0	Oglivie lf	3	1	1	0
L.Smith lf	4	1	1	0	Money dh	4	2	2	0
Iorg dh	4	0	2	1	Moore rf	4	0	1	0
Green pr	0	0	0	0	Gantner 2b	4	1	1	1
McGee cf	4	1	1	0					
O.Smith ss	3	1	1	0					
Total	33	5	8	4	Total	33	7	10	4

St. Louis	....	1 3 0	0 0 1	0 0 0—5
Milwaukee	....	0 0 0	0 1 0	6 0 x—7

E—Gantner, Yount, LaPoint. DP—St. Louis 2, Milwaukee 2. LOB—St. Louis 6, Milwaukee 6. 2B—Oberkfell, Money, L. Smith, Iorg, Gantner. 3B—Oglivie. SB—McGee, Oberkfell. SF—Herr.

	IP.	H.	R.	ER.	BB.	SO.
St. Louis						
LaPoint	6.2	7	4	1	1	3
Bair L,0-1	0	1	2	2	1	0
Kaat	0	1	1	1	1	0
Lahti	1.1	0	0	0	1	0
Milwaukee						
Haas	5.1	7	5	4	2	3
Slaton W,1-0	2	1	0	0	2	1
McClere S,1	1.2	0	0	0	0	2

WP—Haas, Kaat. T—3:04. A—56,560.

Solution to The Other Side of the World (p. 70)

For Milwaukee:

$$PA = 24 \text{ outs} + 7 \text{ runs} + 6 \text{ LOB} = 37$$

Molitor has the last at-bat. Molitor and Oglivie each have a walk and Simmons has two. Here is a table of runs and rbis:

slot	hitter	runs	rbis	positions he can drive in
1	Molitor	1		
2	Yount	1	2	1,6,7,9
3	Cooper	1	1	2,1,7,8,9
4	Simmons	0		
5	Thomas	0	2	3,2,1,9
6	Oglivie	1		
7	Money	2		
8	Moore	0		
9	Gantner	1	1	6,7

What is special is the seventh inning, in which Milwaukee scores 6 runs. According to the Gap Rule, when a batter scores, everybody before him must have already scored or made an out. In particular, Milwaukee's non-scorers—Simmons, Thomas, and Moore—can't all bat before someone who does score in the seventh, because they'd all have to be out first, ending the inning. Thus, for each candidate leadoff hitter, we need to count ahead and make sure we reach 6 scorers before we reach all 3 non-scorers. Using this method, you can see that only Thomas, Oglivie, or Gantner could lead off the inning.

If Gantner leads off the inning, then Money is the sixth run, and at least one more batter, Moore, must bat in the inning. But we proved that Molitor must have Milwaukee's final PA of the game, in the eighth inning, and this would now be impossible.

So either Thomas or Oglivie must lead off the seventh. Next, we see that St. Louis pitcher LaPoint gives up 4 runs (1 in the fifth and 3 in the seventh) and gets two players out in the seventh. But this is impossible if Oglivie leads off, because LaPoint would have to pitch all the way to Simmons in slot 4 to get 2 outs, which means he would pitch to all 6 scorers in the inning, let them all get on base and score, and be charged for all 6 runs. So Thomas must lead off, and LaPoint must get Thomas and Moore out, and must be charged with the runs to Oglivie, Money, and Gantner, accounting for all 3 runs he yields in the seventh. If LaPoint pitched to Molitor as well, he'd similarly be charged with Molitor's run, making him responsible for an additional run, so that's impossible. He pitches only to Thomas through Gantner.

The next pitcher, Bair, gives up 2 runs, yielding a single and a walk and gets nobody out, so he must pitch to Molitor and Yount. They both score in the inning. Bair is responsible for their runs. Bair can't pitch to the next batter, Cooper, because he also scores in the seventh, and Bair would then be responsible for 3 runs. Since Molitor has no hits, he must have walked and Yount must have singled.

The next two pitchers, Kaat and Lahti, also each give up a hit and a walk. Molitor's walk is already accounted for. Kaat pitches at least to Cooper and Simmons, and Simmons doesn't bat again in the game. So Kaat must walk Simmons, and Lahti must walk Oglivie.

Can we determine who drove in which runs? Yes for the last 5 runs in the seventh, but not for the other 2 runs in the game. Money, with 2 runs, had to score the run in the fifth inning. He can't be driven in by Yount or Cooper because, as Thomas leads off the seventh, Yount and Cooper bat in the sixth. Thus, Money is either driven in by Gantner or by no one (Milwaukee has 7 runs but only 6 rbis).

In the seventh inning, when the first 6 batters, Thomas through Molitor, had finished their plate appearances, there were 2 outs (Thomas and Moore) and at most 3 still on base, so by that time Oglivie must have scored, either without an rbi or driven in by Gantner. So now we have 2 runs, one by Money in the fifth and one by Gantner in the seventh, that were either driven in by Gantner or by nobody. We don't know which is which. Gantner has only one rbi. He either drove in Money in the fifth or Oglivie in the seventh. Whomever he didn't drive in scored without an rbi. And in either case the remaining 5 runs must be driven in by the other players with rbis, and in the order in which they bat: Money and Gantner on Yount's single, Molitor by Cooper, and Yount and Cooper by Thomas, all in the seventh inning.

We can fill in some more detail. For Cooper to drive in Molitor, Molitor must have reached base and stayed on base until Cooper batted, so Yount must have batted with the bases loaded: Money, Gantner, and Molitor all on base. And when Thomas drove in Yount and Cooper, the bases must have been loaded also—Simmons must have been on base, too, or else he would have made the third out and the inning would be over.

There's quite a bit we can't figure out. We already showed that Kaat pitches to Cooper and then walks Simmons. Since he got no outs, it seems that he must give up his hit to Cooper, and that Lahti then replaces him, gives up a 2-run single to Thomas, walks Oglivie to load the bases, and gets Money out to end the inning. That's what you'd bet on.

But here's another scenario that fits all the stats in the box score: Cooper drives in Molitor with a ground ball, and reaches first base safely when a fielder tries to throw out Molitor at home. There is no error on the play but it's not ruled a hit. Kaat walks Simmons and gives up a 2 run single to Thomas. Only then does Lahti replace him and walk Oglivie. After that Lahti gets the third out. He could pick off Thomas. Then Money could lead off the eighth inning with a hit.

We also can't pin down exactly when LaPoint makes his error. Since there is no passed ball in the game, LaPoint's error (the only error by St. Louis) must be responsible for all 3 of his unearned runs, so it must have been in the seventh inning. But the error could have been to any of the batters he faced. Here are two possibilities:

(a): Thomas reaches base on an error by LaPoint. Oglivie reaches base on a fielder's choice, with Thomas thrown out at second base. Money singles, and Oglivie goes to second base. Moore is out. Gantner singles, driving in Oglivie. At this point if no error had been made, there would be 3 outs: Thomas, Oglivie, and Moore. The inning would have been over with no runs scored. So all runs given up by LaPoint in the inning are unearned. The run in the fifth inning is scored by Money without any rbi or error. Perhaps he doubles, reaches third base on a ground ball out, is almost picked off third but manages to evade a rundown and score—one of those crazy things that happen once in a while.[2]

(b): Thomas is out. Oglivie singles. Money singles. Moore is out. Gantner reaches base on an error, which also allows Oglivie to score, with no rbi granted. Gantner gets his rbi in the fifth, driving in Money.

Solution to Leary's Bad Day (p. 71); (Hint on p. 171)

All the runs are driven in by the three players who homer, Polonia

---

[2]The runs given up by Bair and Kaat are still earned. As substitute pitchers, the runs they yield can be classified as unearned runs only if the runs would not have been scored but for errors or passed balls that occurred while they were pitching.

(4 rbis), Davis (3 rbis), and White (2 rbis) and they take place in 2 innings, the second (4 runs) and the fifth (5 runs).

Polonia and Parrish each have 2 runs and so score in both the second and fifth innings. Parrish doesn't drive in himself, so in both those innings he's on base when White comes up. White's homer must drive in Parrish (and himself). The only four that Polonia can drive in (and therefore must drive in) are Stevens, Parrish (his other run), Schofield, and himself. Davis is left to drive in Polonia (his other run), Hill, and himself.

With 12 outs, 4 runs, and 0–3 LOB, the first four innings have 16–19 PA. With 15 outs, 9 runs, and 0–3 LOB, the first 5 innings have 24–27 PA. It follows that in the fifth inning Polonia, must bat only as California's 19th batter. It also follows that Stevens and Parrish can only be the 23rd and 24th batters. Hence, Polonia can't drive them in in the fifth; he must do that in the second inning.

In the second inning, Stevens and Parrish must first get on base for Polonia to drive them in. The Gap Rule insists that all intervening players score or make outs. White can't homer (he would drive in Stevens and Parish) and thus can't score (his only run is on his homer), so he is out. Anderson is also out. Thus Schofield must score, driven in by Polonia (as we proved earlier). So the 4 runs of the second inning are Stevens, Parrish, Schofield, and Polonia. We have answered questions 2 and 3: Davis and White must homer in the fifth, since they don't score in the second.

Only 6 of the 9 runs given up by Leary are earned runs. That leaves 3 that are unearned. They must all be due to Leary's fielding error, since there is no other Yankee error or passed ball, and even if Leary's wild pitch contributed to runs scoring, they would still be earned runs.

Could Leary's error be in the second inning? Suppose an error allowed Stevens, Parrish, Schofield, or Polonia to reach base. Then without the error there would have been 3 outs before any runs scored. That would mean that all 4 runs in the inning would be unearned—too many. If instead an error came after Polonia drove in Stevens, Parrish, and Schofield, then the first 3 runs would be earned, so there would be only one unearned run—too few. So there's no way an error in the second inning can account for 3 unearned runs. The error must be in the fifth inning, answering question number 4.

Stevens has to lead off the second; otherwise there would be 3 outs (Winfield, White, and Anderson) before Polonia bats. How did Stevens get on base? Stevens has no hit, walk, or HBP. He can't get on base due to an error, because Leary's error was in the fifth. The only possibility left is that Stevens struck out on a wild pitch and made it safely to first base, so that's what happened. That answers question 5—Leary's wild pitch was in the second inning.

What about question 6? That one is unanswerable. Parrish was hit by a pitch and he also had a walk. These took place in different

Angels 9, Yankees 5

YANKEES	ab	r	h	bi	CALIFORNIA	ab	r	h	bi
Kelly cf	5	2	4	1	Polonia lf	4	2	3	4
Sax 2b	5	1	1	0	DHill 2b	3	1	1	0
Azocar lf	5	0	1	0	CDavis dh	3	1	1	3
Balboni dh	4	0	2	2	Winfield rf	4	0	0	0
Hall ph	1	0	0	0	Stevens 1b	4	1	0	0
Maas 1b	3	0	1	1	Parrish c	2	2	0	0
JeBrfld rf	3	0	0	0	DWhite	3	1	1	2
Leyritz 3b	4	1	1	1	KAnders 3b	3	0	0	0
Geren c	3	1	1	0	Schofield ss	2	1	0	0
Nokes c	1	0	0	0					
Espnoz ss	4	0	1	0					
Total	38	5	12	5	Total	28	9	6	9

```
Yankees ............... 1 0 2   1 0 1   0 0 0—5
California ............. 0 4 0   0 5 0   0 0 x—9
```

E—KAnderson, Leary. DP—Yankees 1. LOB—Yankees 8, California 3. 2B Hits—Kelly, Maas, Polonia. HR—Polonia (2), Leyritz (4), CDavis (12), DWhite (9). SB—Kelly (26). S—DWhite, DHill.

	IP.	H.	R.	ER.	BB.	SO.
Yankees						
Leary L, 6-15	4.2	5	9	6	4	7
JDRobnsn	1.1	1	0	0	1	1
Plunk	2	6	0	0	0	4
California						
Abbott W, 8—10	5.2	11	5	5	0	2
Fraser S, 1	3.1	1	0	0	2	0

HBP—Parrish by Leary. WP—Leary.
Umpires—McCoy, Phillips, Cousins, Hirschbeck.
T—2.37. A—27,937

innings. They have the same effect on events on the field, that is, they can be switched. Thus it is impossible say when the HBP took place.

Only the first question is left. If Jerry is to be believed, it's not possible to say when Polonia hit his homer. To prove this we must show:

One: It's consistent with the box score that Polonia homered in the second.

Two: It's consistent with the box score that Polonia homered in the fifth.

Here we go:

One: Polonia homers in the second inning.

This is relatively easy. Polonia comes up in the second with Stevens, Parrish, and Schofield on base. He hits a grand slam home run. That's the second inning. How then would the runs be scored in the fifth?

We saw earlier that either Schofield, Polonia, Hill, or Davis led off the fifth. We know from the Gap Rule that all five of Polonia, Hill, Davis, Parrish, and White must score before the three non-scorers, Winfield, Stevens, and Schofield, have all come to the plate. The only way to meet both conditions is for Polonia to lead off. Polonia can get on base with his double. Hill can get on base with a hit (or an error). Then Davis can hit a 3-run homer. Winfield and Stevens can make outs. Then Parrish can get on base and White hit a 2-run homer.

What about the three unearned runs? If Hill reaches base on an error, his run is unearned. Polonia's run and Davis' runs are earned. But Parrish and White's runs are unearned since had the error not happened, the inning would have ended with Stevens making the third out.

This scorecard sums all this up:

ANGELS	1	2	3	4	5	6	7	8	9	10	11	ab	R	H	rbi
Polonia	•	HR			2	1						4	2	3	4
DHill	•	1			E	S						3	1	1	0
CDavis	W	•			HR	•						3	1	1	3
Winfield	•			•		•						4	0	0	0
Stevens		WP		•		•						4	1	0	0
Parrish		HB		W		•						2	2	0	0
DWhite		S	•	HR		•						3	1	1	2
K. Anderson		•	•	•		W						3	0	0	0
Schofield		W	•		W	•						2	1	0	0
TOTALS	R H 1	1	4 1		5 3	1						28	9	6	9

Two: Polonia homers in the fifth inning.

Instead of a grand slam in the second inning, Polonia hits his double, driving in 3 runs. Next, Hill hits a single, moving Polonia to third base. Then Davis hits a ground ball to the shortstop while Polonia races home. The shortstop could have thrown to first for the out but tries instead to catch Polonia, to prevent a run. The attempt fails. Polonia scores and Davis is safe at first on a fielder's choice.

Davis doesn't get credit for a hit, since he could have been thrown out but his batted ball does drive in a run so he gets credit for an rbi. There was no error on the play. Even though the shortstop made a mental mistake by not throwing to first base for the third out, mental mistakes aren't scored as errors.

We can keep the same order of at-bats as in *One* as follows: The batter after Davis, Winfield, comes to the plate, but a pickoff throw to first nails Davis, ending the inning.

In the fifth inning Polonia leads off as before, but in this scenario he hits a solo home run. Other than that, the inning looks the same as in the first scenario. The runs by Hill, Parrish, and White are unearned for the same reasons as before.

ANGELS	1	2	3	4	5	6	7	8	9	10	11	ab	R	H	rbi
Polonia		2			HR	1						4	2	3	4
DHill		1			E	S						3	1	1	0
CDavis	W				HR							3	1	1	3
Winfield												4	0	0	0
Stevens		WP										4	1	0	0
Parrish		HBP			W							2	2	0	0
DWhite		S			HR							3	1	1	2
K. Anderson							W					3	0	0	0
Schofield		W			W							2	1	0	0
TOTALS	R:2 H:1	4:1			5:3	1						28	9	6	9

In truth, Polonia did hit a grand slam in the second inning, and it was a special event, an inside-the-park home run. It's a pity there's no way we can ever deduce that detail from the box score!

Solution to Never Out, Never In (p. 73)

Let's say the special guy's place in the lineup was x. Since he ends the game with his walk-off hit, his fourth plate appearance is the last PA for the entire team. So PA is 3 times 9 (three times through the lineup), plus x, the number of batters to get to the special guy the fourth time, which gives us: $PA = 27 + x$.

We can also use our favorite equation: $P = O + R + LOB$. Since it's a walk-off win, the game has at least 9 innings. Since the walk-off comes with 2 outs in the final inning, $O \geq 26$. The team scores 6 runs. Since the special guy was never out and never scored, he has 4 LOB, so for the whole team LOB ≥ 4. So PA $\geq 26 + 6 + 4$, which reduces to PA $>= 36$. Combining the 2 PA equations, $27 + x$ must be 36 or more, so x must be 9 or more. But x can't be more than 9, since there are only 9 slots in the lineup. The special guy must bat in slot 9, the final slot.

Solution to Double Plays (p. 74); (Hint on p. 171)

Foster can drive in a run if he is third or fourth to bat in an inning. He can't come up as the fifth batter in an inning because with 5 batters, the inning will either have 2 runs or a run and a man left on base—both of which violate the conditions of the puzzle.

If Foster drives in a run as the fourth batter, then the next time he is up he will be the third batter (the intervening innings have no runs or LOB, so have only 3 PA). But after driving in a run as the third batter, he will next come up as the second batter and can no longer drive in a run. Consequently, Foster can't drive in a run more than twice. But we are told that he did drive in a run at least twice. Thus, since Foster drives in all the Blackbirds' runs, the Birds score a total of 2 runs.

The Birds scored 2 runs in total, but only one run per scoring inning. So they scored 1 run in the final inning, an extra inning, and at the end of the previous inning, they had 1 run. The score must have been tied 1-1 then, because otherwise the game would have been over. So the final score must be 2-1.

Solution to Mistakes Were Made (p. 75); (Hint on p. 171)

With 4 runs, 27 outs, and 5 LOB, the Cats have 36 PA, so Ingals in slot 9 was their last batter, and he hit into a double play. Each batting slot had 4 plate appearances ($4 \times 9 = 36$); thus Brown, Frank, and Guzman, with 3 at-bats each, must have received the 3 walks given up by Dogs' pitching.

Engel and Brown have homers. Neither homer was in the ninth inning, because the Dogs' pitcher in the ninth, Ichikawa, gave up no hits, and thus no homers. Engel can't homer in the first inning, because he bats in slot 5, and at least two of the preceding batters would have had to score, making 3 runs, when there were only 2. So Brown must homer in the first, and Engel in the seventh.

There were three errors committed by Dogs' fielders. The three pitchers for the Dogs each gave up 1 unearned run. It makes sense to look at them individually.

Ignacio pitched the seventh inning, and the run charged to him is unearned. There's only one way Engel's homer (or any homer) can be an unearned run, and that's if errorless play would have resulted in 3 outs before the homer was hit. But any players who batted before Engel must have been out before Engel scored—otherwise there would have been at least 2 runs in the seventh, not one. There were at most 2 previous batters up. No matter whether any errors had been made on them, 2 batters can at most make 2 outs. There's no way that errorless play could have resulted in 3 outs before Engel reached the plate.

So does this mean there's a mistake in the box score, that Engel's run must have been earned? No, there's one more possibility, which is: before Engel hit his homer, he hit a foul fly ball that should have been caught, but was dropped for an error. Since Engel should have been out, the run from his homer would be unearned. That's what must have happened.

Ito pitched at the start of the game. Recall that Brown's homer was in the first inning. The other

Cats	ab	r	h	bi	Dogs	ab	r	h	bi
Abe ss	4	1	0	0	Amsler 1b	4	1	1	0
Brown 3b	3	1	2	1	Blanco cf	3	1	1	1
Castro cf	4	0	0	0	Clark lf	4	0	0	0
Dominguez c	4	1	0	0	Durand rf	4	1	1	0
Engel rf	4	1	1	1	Emerson 2b	4	0	1	1
Frank 1b	3	0	1	0	Foster 3b	3	0	0	1
Guzman lf	3	0	0	0	Gato c	3	0	0	0
Hall 2b	4	0	0	1	Herrera ss	4	0	1	0
Iglesias p	2	0	0	0	Ito p	2	0	0	0
Ivory ph	1	0	1	0	Ignacio p	1	0	0	0
Ingals p	1	0	0	0	Ichikawa p	1	0	0	0
Totals	33	4	5	3	Totals	33	3	5	3

```
Cats  2 0 0   0 0 0   1 0 1--4
Dogs  0 0 0   0 0 0   0 0 3--3
```

	ip	h	r	er	bb	so
Iglesias (W)	7	2	0	0	1	5
Ingals	2	3	3	3	2	2
Ito	6	2	2	1	0	1
Ignacio	2	3	1	0	1	0
Ichikawa (L)	1	0	1	0	2	2

LOB--Cats 5, Dogs 6, HR--Brown, Engel, Foster, 2b--Frank, E--Blanco, Gato, Herrera, GIDP--Ingals

run in the first must be scored by Abe. (The only alternative is that Abe scores in the ninth, but that would require 9 batters and 3 runs in that inning in order to bat all the way to Ingals.) But how did Abe get on base as leadoff hitter when he had no hits, walks, and wasn't hit by a pitch? Not through a strikeout combined with a wild pitch or passed ball, because there are no WP or PB noted in the box score. He must have reached base with the help of an error. This error also explains why there is only one earned run given up by Ito, leaving the other run, namely Abe's, to be unearned.

Finally, Ichikawa pitched the ninth inning. He has credit for 2 strikeouts. But Ingals, the Cat's pitcher, who was the last batter in the game, hit into a double play. There can't be 4 outs in the inning! The only resolution to this problem is that one of the strikeouts didn't cause an out. Instead the ball wasn't caught by the catcher and the batter reached base, on a WP, PB, or error. But it can't be from a WP or PB, so it must be on the third error of the game. Since all the other runs are accounted for, the run in the ninth was scored by Dominguez.

There were at most six men up in the ninth: 1 to score the run, 3 who made outs, and at most 2 LOB (because of the game-ending double play). So Dominguez led off the inning. But how did he reach base? He couldn't reach base on a fielder's choice because no one was on base. He had no hits, no HBP, and no walks. And he can't reach base on a dropped third strike that was ruled to be due a wild pitch or a passed ball, because there are no PB or WP in the box score. The only remaining possibility is some sort of error.

There are only 3 errors in the game. So there must be just one error in each of the three scoring innings. Thus, the error that allowed Dominguez to reach base must be the same error that allowed a batter in the ninth to reach base on a strikeout—in other words, we have proven that it was Dominguez who must have reached base on an error after the catcher dropped the third strike. This could have happened without a passed ball or wild pitch being called if the official scorer ruled that the catcher could have thrown Dominguez out at first after dropping the ball, but made a wild throw to first, and this error was the reason he was safe.

To summarize, we have accounted for all three errors. The first was made during Abe's plate appearance in the first inning, allowing him to reach base and score an unearned run before Brown's homer. The second was made during Engel's plate appearance in the seventh, and we proved that a fielder must have dropped a foul fly ball that otherwise would have ended the inning before he could hit his homer. The third was made when Dominguez led off the ninth inning and allowed him to reach first base even though he had just struck out.

We can actually reconstruct the Cats' entire ninth inning. Ichikawa gave up 2 walks, which must have been to Frank and Guzman. Hall has an rbi, so he certainly didn't strike out. The second strikeout by Ito must have been Engel. So, in order: Dominguez struck out but reached first on an error. Engel struck out. Frank and Guzman walked, loading the bases. Hall, with no hits, reached base on a fielder's choice, driving in Dominguez. Ingals was allowed to bat with the bases still loaded, but hit into an inning-ending double play. He then he nearly blew the game by giving up 3 runs in the bottom of the ninth.

It is tempting to try to figure out which fielder made which of the errors. But we must recognize that this is just speculation, not rigorous logical deduction. The rules of baseball don't specify where any fielders, other than the pitcher and catcher, can play on the field. So there is no way to logically prove which fielder made which of the errors.

Still, we can make some reasonable guesses. Engel's foul fly ball probably was out of reach of the centerfielder, Blanco, because centerfielders almost always play very far from the foul line. And there's no plausible way he could have made the error allowing Dominguez to reach base on a dropped third strike. So we conclude that Blanco's error was (probably) in the first inning. He dropped a fly ball by Abe, allowing him to reach base and score. And the shortstop, Herrara, was also too far from the action for an error of his to let Dominguez reach base on a strikeout, so he was the one who (probably) dropped the foul pop fly by Engel. This leaves the catcher, Gato, to have the error in the ninth. He could have either made a bad throw to first or dropped the ball a second time when he was trying to tag Dominguez out.

The scorecard below shows how the game could have transpired. The 'K' is for a strikeout.

Team I	1	2	3	4	5	6	7	8	9	10	11	ab	R	H	rbi
Abe	◇		•		•		•					4	1	0	0
Brown	HR		1			•		w				3	1	2	1
Castro	•			•			•	•				4	0	0	0
Dominguez	•			•			•		K			4	1	0	0
Engel	•			•				HR	•			4	0	1	1
Frank		•			•		2	w				3	1	1	0
Guzman		•			•		•	w				3	0	0	0
Hall		•			•			•	1			4	0	1	1
Iglesias			•			•						2	0	0	0
Ivory								1				1	0	0	0
Ingals									DP			1	0	0	0
TOTALS	R											33	4	5	3

—where the errors are

Note that the run in the ninth was unearned, because runs scored by any player who reaches base on an error are always unearned.

Solution to The Two Rods (p. 76)

Since GRod came up four times and was the last batter for Anytown, Anytown had exactly 33 plate appearances. Since the problem specifies that GRod "went down on strikes every time," it means that every time he struck out, the catcher successfully caught the ball, putting him out, and he was not able to run safely to first base. GRod's strikeouts with the bases loaded mean there were at least 6 men left on base. If Anytown played a full 9 innings, there would be 27 outs; hence no runs for Anytown. We're given, though, that both teams scored, so Anytown played fewer than 9 innings. Anytown couldn't have won in a ninth inning walk-off, since GRod's last plate appearance, a strikeout, ended the inning. The strikeout must have ended the eighth inning. So Anytown won as the home team, ahead after 8 1/2 innings. With 24 outs then, they could have scored as many as 3 runs.

Anytown	1	2	3	4	5	6	7	8	9	10	11	ab	R	H	rbi
	•		•			•	(w)					3	1	0	0
	•		•			•	HR					4	1	1	3
	1b		•			1b	•					4	0	2	0
	w			•		1b		•				3	0	1	0
	w			•		w		•				2	0	0	0
GRod	K			K		K	K					4	0	0	0
		•			•		•					3	0	0	0
		•			•		•					3	0	0	0
		•			•		(w)					2	1	0	0
TOTALS	R/H	1					2	3/1				28	3	4	3

The same isn't true of Otherburg. Since HRod was seventh in the lineup, Otherburg had 34 plate appearances in 9 innings. With 27 outs, Otherburg must have scored exactly 1 run with 6 LOB, for example:

OtherBurG	1	2	3	4	5	6	7	8	9	10	11	ab	R	H	rbi
	•	•			•		•					4	0	0	0
	•		•			•		•				4	0	0	0
	HR		•			•		•				4	1	1	1
	1b		•			1b		•				4	0	2	0
	w			•		w			•			2	0	0	0
	w			•		w			•			2	0	0	0
HRod	K			K		K			K			4	0	0	0
		•			•		•					3	0	0	0
		•			•		•					3	0	0	0
TOTALS	R 1 H 1	1					1					30	1	3	1

Solution to The Guy Who Did it All (p. 77)

First, Farrell can't score twice in an inning (see the Gap Rules, p.24). That means they can't win 6-4 in extra innings. That also means that after 8 1/2 innings the Blue Sox must have at least 5 runs to its opponent's 4 and so the game is over. Since it's over, the Blues must actually have all 6 of its runs in the first 8 innings.

Second, Farrell *can* score in 2 consecutive innings. If he leads off an inning and scores and the inning has the maximum 3 LOB, 7 batters, Farrell will be third up in the following inning and can score again.

Third, Farrell *can't* score in 3 consecutive innings. To score, he must be among the first 3 batters (Gap Rule again). If Farrell doesn't lead off an inning, then he won't be one of the first 3 batters in the next inning.

This leaves us with just one possible line score: 110 110 11. Any other arrangement with two 0's would have 3 scoring innings in a row. So we must have Farrell lead off the seventh inning, and we must have 7 batters in that inning, leaving the bases loaded. That answers the question.

There's a nice symmetry here: the scoring pattern of 110 110 11 is one symmetry, and the fact that we can perfectly repeat the sequence of the first innings indefinitely is another form of symmetry. And it's pleasing that it just barely works.

Solution to Scraps (p. 78)

	aB	r		ip	h	r	er	BB	SO
Blanco cf	5	1	Faatz	1 1/3	1	2	2	0	1
Duffy 1B	4	1	Jackson	0	1	2	2	1	0
Martin p	4	1	Kimura	6 2/3	9	4	4	0	3
Namath 3B	5	1	1	2					
Okada ss	4	1	1	3					
RadBourn c	4	0	1	0					
Shugart lf	3	1	1	0					
Smith 2B	5	1	2	0					
Suarez rf	5	1	2	2					

Westwood	4	0	0	0	0	0	0	0	2	—6
Eastham	3	0	0	0	0	0	3	0	2	—8

LOB—Eastham 9, Westwood 0, HR—Okada, Namath,
HBP—Duffy By Faatz

The picture so far:

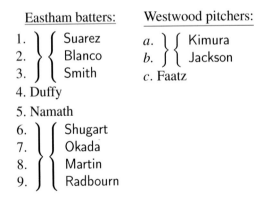

Eastham batters:

1. ⎫
2. ⎬ { Suarez / Blanco / Smith
3. ⎭
4. Duffy
5. Namath
6. ⎫
7. ⎬ { Shugart / Okada / Martin / Radbourn
8. ⎭
9.

Westwood pitchers:

a. ⎫ { Kimura / Jackson
b. ⎭
c. Faatz

Could Jackson be the starting pitcher? He can't get anybody out (he has 0 ip), and Shugart, whom he walks, can't bat until at least slot 6. The leadoff batter could get the one hit Jackson yields. After that, players could reach base on fielder's choices. But after 6 slots batted, there would still be no outs, at most 3 on base, and thus at least 3 runs scored, all charged to him, when he really gives up only 2 runs. So he's not the starter. The order of pitchers is Kimura, Jackson, Faatz.

Now we'll sort out the batters, in particular who scores when. To get started, recall that every Eastham batter has a single run except Radbourn, who has none. And the 4 and 5 batters, Duffy and Namath, scored their runs in the ninth, so they can't score in other innings.

Eastham scored 3 runs in the first and 3 in the seventh. It must be that the batters in slots 1, 2, and 3 scored in the first. If not, then someone batting sixth or later would have to score in the first. But before anyone batting 6 or later scores, batters 1, 4, and 5 would also have to be out, which is impossible. Thus Suarez, Blanco, and Smith scored in the first and Shugart, Okada, and Martin scored in the seventh inning.

Since Kimura and Jackson gave up the first 6 runs of the game, they must have pitched to scorers Shugart, Okada, and Martin in the seventh. A subsequent batter, who can only be Radbourn, must have made the final out to end the inning. (The batters in slots 1-3 must bat in the eighth inning, so they can't bat in the seventh.) That forces Radbourn to be in slot 9 and Shugart, Okada, and Martin, in some order, to be in slots 6-8. Faatz must come in to pitch to Radbourn in order to get his 1 1/3 innings pitched.

The runs by Shugart, Okada, and Martin had to be driven in by somebody in slots 6-9. But the only one of them with any rbis is Okada, so he must have driven in all 3 runs, including his own. The only way to do this would be on a homer, so he had a 3-run homer in the seventh. The only way this could happen is if Shugart and Martin, in some order, bat in slots 6 and 7, and Okada bats in slot 8. So the bases were empty when Radbourn batted and made the final out.

Jackson was responsible for the final two runs of the seventh inning. So he must have pitched to Okada and to the previous batter, who must have been Shugart, the batter he walked. Kimura, the previous pitcher, must have pitched to Martin in order to be responsible for his fourth run. That places Martin in slot 6 and Shugart in slot 7.

To finish the problem, we need to order the first 3 batters: Suarez, Blanco, and Smith, each of which scored a run. Since all other rbis are accounted for, two of these batters were driven in by Suarez and one by Duffy, batting fourth. Thus, Suarez, who didn't have a home run,

must have driven in two previous batters, namely Blanco and Smith.
Duffy must have driven in Suarez. So Suarez was in slot 3. Blanco
can't be the leadoff batter, because he has no hits, walks, or any other
way to get on base (there are no errors in the game, and no WP or
PB). So Smith must have led off with a single, and Blanco must have
batted second and reached base on a fielder's choice. We now have
both the batting and pitching orders completely unscrambled:

	aB	r		ip	h	r	er	BB	so
Smith 2B	5	1	Kimura	6 2/3	4	4	4	0	3
Blanco cf	5	1	Jackson	0	1	2	2	1	0
Suarez rf	5	1	Faatz	1 1/3	1	2	2	0	1
Duffy 1B	4	1	2	1					
Namath 3B	5	1	1	2					
Martin p	4	1	1	0					
Shugart lf	3	1	1	0					
Okada ss	4	1	1	3					
Radbourn c	4	0	1	0					

Solution to At the Dawn of Time (p. 85); (Hint on p. 172)

1. Which is the correct meaning?
Two possible meanings for outs listed in a player's slot:

(A) Somebody was put out during this player's PA.

(B) This player was personally put out sometime.

First, note that with
9 outs, 11 runs, and a maximum
of 9 LOB (3 each inning) there
are at most 29 PA in the game.
Now let's see what meaning (B)
would imply for Curry's Team.
If (B) is the rule, then Curry
personally made outs in 3 plate

	HANDS OUT			**RUNS**
CURRY'S TEAM				
	1	2	3	
Curry	o1	o2/o3		xx
Niebuhr			o1	xxx
Maltby			o3	x
Dupignac	o2			xx
Turney		o1		xx
Clare			o2	
Gourlie	o3			x
Total				**11**

appearances and scored runs in 2 other plate appearances, making at least 5 PA. That would require going around the batting order four times, for 28 PA, and a 29th PA for Curry to bat again, after which Maltby would have to make the third out—at least 31 PA. So Rule B requires more than the maximum 29 PA—it's impossible. Rule A must be the correct meaning here. "o2" at player X's location, for example, isn't, strictly speaking, an out by X. Call it a "charged out," meaning that someone was put out on the play and it was X's fault—i.e, X's batting resulted in the out.

2. What is o2/o3?

Under meaning (A), o3 next to a slot means that slot had the last PA of the inning. So, Gourlie had the last PA of the first inning. The first out of the second inning occurred during Turney's slot—the 12th of the game. If Curry was up twice in the second inning after Turney, once for o2 and once for o3, that would reach the 28th PA of the game. The whole game would take way more than the maxiumum 29. So the o2/o3 here means a double play—2 outs on Curry's third plate appearance.

3. Who scored when for Curry's team?

From the discussion above, we know the shape of each inning's plate appearances. →

	1	2	3
Curry	☐	☐ ☐	☐
Niebuhr	☐	☐	☐
Maltby	☐	☐	☐
Dupignac	☐	☐	☐
Turney	☐	☐	☐
Clare	☐	☐	☐
Gourlie	☐	☐	☐

We know, furthermore, that a player leading off with "o1" can't score in that at-bat, and the last player in an inning also can't score. Finally, from the original box score we know that Clare scores no runs.

We can place a dot where we know a player can't score. That gives us this picture →

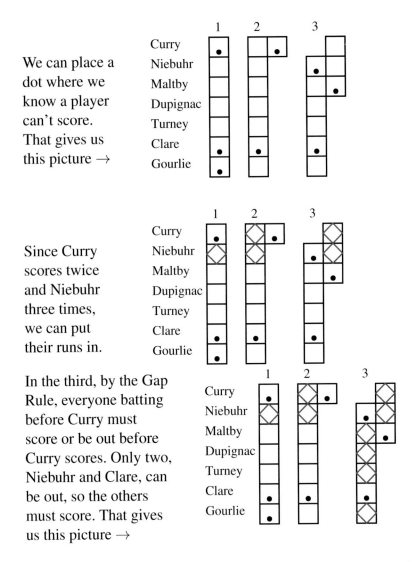

Since Curry scores twice and Niebuhr three times, we can put their runs in.

In the third, by the Gap Rule, everyone batting before Curry must score or be out before Curry scores. Only two, Niebuhr and Clare, can be out, so the others must score. That gives us this picture →

Dupignac and Turney each score an additional run. These can be in the first or the second, except that if Turney scores in the first, Dupignac must as well (otherwise Maltby and Dupignac would have to be out, ending the inning).

On the chart we can now place where every charged out lies. This is easy in the case of the first inning. In the second, note that Curry can't have caused an out leading off and scoring, so he must have hit into a double play in his second at-bat of the inning. In the third inning, in addition to placing each charged out, we can conclude that since every other batter scores, the players with charged outs were personally put out as opposed to causing a different player to be put out.

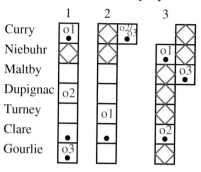

Solution to Later That Day (p. 88)

Since Brodhead causes the last out, o3, of the last inning, the total number of PA must be a multiple of 7. With 8 runs and 9 outs, there must be either 4 LOB to make 21 PA or 11 LOB to make 28 PA. But 11 LOB is impossible in 3 innings, so the team had 21 PA, batting three times around the order.

	HANDS OUT			RUNS
CARTWRIGHT'S TEAM				
	1	**2**	**3**	
Cartwright	o1	o2		x
Moncrief	o2	o3		x
De Witt	o3			xx
Tucker				xxx
Smith		o1	o1	
Birney				
Brodhead			o2/o3	x
Total				8

Minimal numbers for the plate appearances of the 3 innings that will place the three charged outs are 3, 6, and 5. That clearly does not provide enough for 21 PA.

One inning must have 7 more PA for a total of 21. The additional 7 PA can't be in the first inning from what Jim read on the website. Can they be in the second?

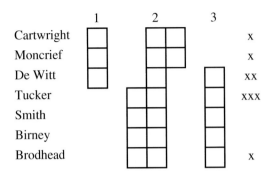

They can't. With 13 PA in the second inning there will be 7 runs, leaving only 1 run for the third inning. But De Witt must score 2 runs and Tucker must score 3, so both must score in the third inning. And that's not possible.

The picture, then, is this, with Moncrief causing the third out of the second.

	1	2	3	
Cartwright	☐	☐	☐	x
Moncrief	☐	☐	☐	x
De Witt	☐		☐	xx
Tucker		☐	☐	xxx
Smith		☐	☐	
Birney		☐	☐	
Brodhead		☐	☐	x

As with the Curry team, we can place dots to indicate plate appearances where runs can't be scored—the ends of innings, all 3 PA in the first inning, and all the PA for Smith and Birney, who don't score.

	1	2	3	
Cartwright	•	☐	☐	x
Moncrief	•	•	☐	x
De Witt	•		☐	xx
Tucker		☐	☐	xxx
Smith		•	• •	
Birney		•	• •	
Brodhead		☐	•	x

Then we can place some runs where they must go—the 2 for De Witt, the 3 for Tucker, and the 1 for Moncrief.

In the third inning, the Gap Rule says that before Moncrief scores, at most 2 batters can be out and the rest must score. Smith and Birney can't score, so they must be out. The other two players, Brodhead and Cartwright, must score, so in they go.

Now we place the charged outs. Note that in this case, "o2/o3" isn't a double play. There must be 2 outs between Tucker's first run and Brodhead's run.

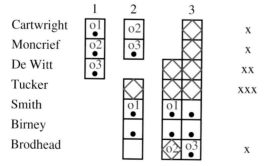

You might wonder how Brodhead scores with his "o2." But this is what must have happened: Birney got on base. Then Brodhead reached base on a fielder's choice, with Birney out on the play. Birney's out is charged to Brodhead.

Solution to Ancient History (p. 88)

Can we say who was catching for Cleveland in the ninth inning?

Myatt has an at-bat. If he batted in the eighth inning, then since he's the last catcher listed in the box score, he had to be the catcher in the ninth when Edwards was pitching.

The PA equation for Cleveland, PA = 27 outs + 1 run + 11 LOB = 39, tells us that the last batter for Cleveland was Speaker. The box score tells us that Gardner batted for Coveleskie in the eighth inning. Since slot 9 batted in the eighth, if Myatt batted in the ninth, it must be 8 slots after Gardner's pinch hit, and Cleveland would have had to bat around to Speaker in slot 3 to finish the inning. That makes at least 13 batters in the eighth and ninth innings. But these innings have 6 outs and and at most 6 LOB, so Cleveland would have had to score a run, which they didn't. So Myatt must have batted in the eighth, and was the catcher in the ninth.

Can we conclude that a third strike was dropped in the ninth?

Edwards had a strikeout, but the catcher, Myatt, had no putout. Thus, he didn't catch the third strike.

Can we conclude that the runner was out at first?

No, we can't. Myatt may have thrown the ball to first base, and the first baseman, Brower, may have muffed the catch, allowing Edwards to be safe at first. After all, Brower does have an error. On the other hand, Myatt has an assist, so it's possible he threw to first and the batter was out. We know that Myatt didn't tag him out, because that would have been recorded as a putout for Myatt. In any case, Myatt was not charged with an error or a passed ball, so if the batter reached first base safely, the official scorer didn't deem it to be his fault.

WASHINGTON (A.)	Ab	R	H	Po	A	CLEVELAND (A.)	Ab	R	H	Po	A
Leibold,cf	4	1	0	2	0	Jamieson,lf	5	0	3	0	
Bush,2b	5	0	1	2	3	Summa,rf	5	1	1	1	
Goslin,lf	4	0	0	1	0	Speaker,cf	3	0	2	0	0
Rice,rf	4	1	1	2	0	J. Sewell,ss	3	0	1	3	4
Ruel,c	5	0	2	8	0	Wamb'ss,2b	4	0	1	4	4
Hargrave,3b	3	0	0	0	1	Lutzke,3b	3	0	1	0	3
Peck'ugh,ss	2	0	0	3	2	Brower,1b	4	0	2	13	0
Evans,1b	3	1	2	8	1	O'Neill,c	2	0	0	3	1
Johnson,p	2	0	0	1	0	Myatt,c	1	0	0	1	
						Coveleskie,p	3	0	0	0	3
Total	35	3	6	27	9	Edwards,p	0	0	0	0	0
						aGardner	1	0	0	0	0
						Total	34	1	8	27	17

aBatted for Coveleskie in eighth.
Errors — Washington 0, Cleveland 3 (J. Sewell, Brower, O'Neill).

Washington............. 1 0 1 0 0 0 1 0 0—3
Cleveland............. 0 0 1 0 0 0 0 0 0—1

Two-base hits — Wambsganss, Summa, Brower (2). Home run—Rice. Sacrifices—Bush, Johnson, Lutzke. Double plays—Wambsganss, J. Sewell and Brower; J. Sewell, Wambsganss and Brower; Summa and Brower; Coveleskie, Wambsganss and Brower. Left on bases—Washington 1, Cleveland 11. Bases on balls—Off Johnson 4, Coveleskie 1. Hits—Off Coveleskie 6 in 8 innings, Edwards 0 in 1. Struck out—By Johnson 5, Coveleskie 1, Edwards 1. Losing pitcher — Coveleskie. Umpires — Hildebrand and Owens. Time of game—1:40.

for Myatt. In any case, Myatt was not charged with an error or a passed ball, so if the batter reached first base safely, the official scorer didn't deem it to be his fault.

Can we locate the inning when Rice hit his homer?

Rice, batting in slot 4, must bat in either the first or second inning. After 3 innings, there are 9 outs, 1 run, and at most 1 LOB (Washington has only 1 LOB in the whole game), for at most 11 PA. So Rice can't bat a second time in the third inning. So Rice's homer must be in the seventh. Evans' run comes in the third.

Solution to Baseball at War (p. 91)

The key is that Chicago left only one man on base. That means we have a pretty good idea of when the innings start and end. For example, there are no runs in the first inning so there will be 3 players up (if no one is left on base) or 4 (if this is the inning with the one man left on).

```
                         FIRST GAME
          CHICAGO (N.)              NEW YORK (N.)

            a.b.r.h. po.a e.                  a.b.r. h. po.a. e.
Hack. 3b .... 4   0 0 1   4 0   Rucker. cf ... 5   0 0 3   1 0
Hughes. ss... 3   1 0 2   2 0   Hausm'n. 3b. 4    1 1 2   6 0
Car'r'ta. 1b.. 4  1 1 12  0 0   Ott. rf....... 3   2 2 0   0 0
Nich'ls'n. rf. 3  4 3 2   0 1   Medwick. lf.. 4   0 1 0   0 0
Dal's'dro. lf.. 4 0 1 1   0 0   Weint'b. 1b.. 4   0 1 15  0 0
Pafko. cf..... 4  0 0 0   0 0   Lombardi. c. 3    0 1 4   1 0
Johnson. 2b . 4   1 3 4   4 0   Kerr. ss ..... 4   1 0 1   5 0
Kreitner. c .. 3  0 0 5   0 0   Luby. 3b..... 3    0 0 2   3 0
Chipman. p . 2    0 0 0   0 0   Voiselle. p ... 1  0 1 0   1 0
V'and'b'g. p. 1   0 1 0   0 0   Adams.. p... 0    0 0 0   0 0
                                aSloan........ 1   0 0 0   0 0
Total         32  7 9 27 10 1   Hansen. p... 0    0 0 0   0 0
                                bGardella.... 1   0 0 0   0 0

                                Total         33   4 7 27 17 0
aBatted for Adams in the seventh.
bBatted for Hansen in the ninth.

Chicago ...........................010  111  010—7
New York ...........................000  111  010—4
```

Runs batted in—Johnson, Nicholson 4, Ott, Voiselle, Cavar-retta, Medwick, Vandenberg, Weintraub. Two-base hit—Weintraub. Home runs—Nicholson 3, Ott. Stolen base—Hughes. Double plays—Hausmann, Kerr and Weintraub; Hack, Johnson and Cararretta; Rucker, Luby and Hausmann. Left on bases—New York 6, Chicago 1. Bases on Balls—Off Voiselle 3, Chipman 1, Vandenberg 2. Struck out—By Voiselle 2, Chipman 1, Hansen 1, Vandenberg 1. Hits—off Chipman 6 in 5 1–3 innings, Vandenberg 1 in 3 2–3 innings, Voiselle 7 in 6 (none out in seventh), Adams 1 in 1, Hansen 1 in 2. Hit by pitcher—By Vandenberg (Luby). Wild pitch—Chipman. Winning pitcher—Chipman. Losing pitcher—Voiselle. Umpires—Conlan, Barr and Sears. Time of game—2:10.

Then the second inning either starts with slot 4 or 5 and runs to slot 7 (if no one is left on base in the first 2 innings) or slot 8 (if someone is left on base in one of the first 2 innings). We get this picture of the innings where we know ±1 where each inning begins and ends, depending on when Chicago leaves a man on base:

Chicago	1	2	3	4	5	6	7	8	9	10	11	ab	R	H	rbi
Hack	•				•	?	?					4	0	0	0
Hughes	•		?	?		•						3	1	0	0
Cavaretta	•			•		•		•				4	1	1	1
Nicholson	?	?		•		•		•				3	4	3	4
d'alessandro		•		•		•		?	?			4	0	1	0
Pafko		•		?	?	?	?		•			4	0	0	0
Johnson		•			•		•		•			4	1	3	1
Kreitner		?	?		•		•		•			3	0	0	0
Chipman			•		?	?						2	0	0	0
Vandenberg						•						1	0	1	1
TOTALS	R		1		1		3	1	1			32	7	9	7
	H														

Nicholson scored four times. Those must be in innings 2, 4, 6, and 8, the only 4 innings when Nicholson batted and runs were scored.

Who scored in the seventh? The only possibility is Johnson. That leaves Hughes and Cavaretta to score in the sixth.

Nicholson hit 3 home runs (driving himself in three times) and in one of those 4 innings, he was driven in by ... whom? He can only be driven in by Johnson or Vandenberg (by the Five-step Rule). But from the structure of the innings, it has to be in the second inning and it has to be Johnson. Since Nicholson has 3 at-bats, he must have one of the walks. That was in the second, the only time he didn't homer.

Who did Vandenberg drive in? The only possibility is Johnson. Finally, we must have that Cavaretta drove in Hughes and Nicholson drove in Cavaretta.

We can resolve where most of the innings start and finish. Nicholson's run means that the second starts with him. Nicholson's eighth inning homer can't be preceded by 3 batters (who would have to be out) so the eighth begins with Hughes and consequently the ninth begins with Pafko.

Chicago	1	2	3	4	5	6	7	8	9	10	11	ab	R	H	rbi
Hack	•		•			•	•					4	0	0	0
Hughes	•		?	?		•		•				3	1	0	0
Cavaretta	•		•			•		•				4	1	1	1
Nicholson		W		HR		HR		HR				3	4	3	4
d'alessandro		•		•		•		•				4	0	1	0
Pafko		•		?	?	?	?		•			4	0	0	0
Johnson		•			•		•		•			4	1	3	1
Kreitner			?	?	•		•		•			3	0	0	0
Chipman			•		?	?						2	0	0	0
Vandenberg						•						1	0	1	1
TOTALS	R/H		1/		1/		3/	1/	1/			32	7	9	7

Solution to **Twins Fall Short in Slugfest** (p. 92); (Hint on p. 172)

Doing the math, Minnesota had 46 plate appearances so the first slot had 6 PA and all the others had 5 PA. We can tell, then, that Battey and Zimmerman shared slot 6 and that all the guys from

Kralick on down shared slot 9. Further, Rollins, Killebrew, Allison, Allen, and Versalles all had walks.

Mincher pinch-hit for Gomez. He just had that 1 plate appearance but he made it count for 3 rbis. Since he had a homer, it was a 3-run homer, and must have been in either the sixth or the ninth inning. But it couldn't be in the ninth inning, because there were 2 more plate appearances in slot 9 after Mincher. So it was in the sixth inning. [Note: we don't need the footnote in the box score to figure this out!]

Bunning, pitching for Detroit, is listed as pitching 5 innings, but he gave up 6 runs, so he must have pitched into the sixth and surrendered Mincher's homer. He recorded no outs in the sixth, so Mincher's homer came with nobody out. Thus, Allen led off the sixth and Mincher drove in Allen and Versalles with his homer.

The box score lists each team's homers in the order they appeared, first the homers of Minnesota, the visiting team, then Detroit's homers, so the third homer, which came after Mincher's, was by Allen. The only subsequent inning with any runs was the ninth inning, so Allen hit his homer in the ninth.

With 46 plate appearances for Minnesota the last batter in the ninth was Green. Since Green and Versalles had no rbis, and since the ninth slot's rbis were in the sixth, and since all 9 runs had corresponding rbis, there can't be any runs in the ninth after Allen's homer.

Who besides Allen scored in the ninth? Not Power, because between Allen and Power were 4 batters, only one of whom could score—the other three would be outs. So Allison and Zimmerman scored the other 2 runs. We know Allison didn't homer in the

MINNESOTA (A.)	ab.	r.	h.	rbi
Green, cf	6	0	2	0
Power, 1b	5	1	2	0
Rollins, 3b	4	0	1	1
Killebrew, lf	4	0	0	0
Allison, rf	4	2	1	2
Battey, c	3	0	1	0
Zim'man, c	2	1	2	0
Allen, 2b	4	2	1	3
Versalles, ss	4	1	2	0
Kralick, p	2	1	1	0
Gomez, p	0	0	0	0
bMincher	1	1	1	3
Stange, p	0	0	0	0
Pleis, p	1	0	1	0
Sullivan, p	0	0	0	0
cOliva	1	0	0	0
Total	41	9	15	9

DETROIT (A.)	ab.	r.	h.	rbi
Fernandez, ss	4	3	3	1
Bruton, cf	4	2	2	1
Kaline, rf	3	1	2	3
Colavito, lf	5	0	2	2
Morton, 1b	2	1	1	1
aCash, 1b	3	0	1	1
Kostro, 3b	4	0	1	0
McAuliffe, 2b	3	1	1	1
Brown, c	5	0	1	0
Bunning, p	3	1	1	0
Fox, p	1	1	0	0
Humphreys, p	0	0	0	0
Total	37	10	15	10

aSingled for Morton in 5th; bHit a homer for Gomez in 6th; cStruck out for Sullivan in 9th.

Minnesota	0 0 1	0 2 3	0 0 3—	9
Detroit	0 1 1	0 3 1	4 0.—	10

E—None. A—Minnesota 9, Detroit 14. LOB—Minnesota 10, Detroit 11. 2-B Hits—Power, Cralick, Rollins, Zimmerman, Kaline 2, Fernandez. HR—Allison, Mincher, Allen, Morton, McAuliffe. SB—Fernandez. Sacrifices — Bruton, Kostro. SF — Kaline.

	IP.	H.	R.	ER.	BB.	SO.
*Kralick	4	7	5	5	2	0
Gomez	1	1	0	0	1	1
Stange	2⅓	2	1	1	0	1
Pleis (L, 2–5)	1	3	4	4	1	1
Sullivan	1⅓	2	0	0	1	0
†Bunning	5	11	6	6	2	3
‡Fox (W, 2–1)	3	3	2	2	3	0
Humphreys	1	1	1	1	0	1

*Faced 4 batters in 5th. †Faced 3 batters in 6th. ‡Faced 2 batters in 9th.

Wild pitch—Bunning. Balk—Fox. Umpires—Hurley, Flaherty, Runge, Carrigan. Time of game—3:26. Attendance—10,137.

ninth, because his homer is listed before Mincher's, so it occurred before Mincher's homer in the sixth. The only remaining possibility is that Allen drove in all the runs of the ninth with a three-run homer.

We've settled the sixth and the ninth innings. We still have runs by Allen, Power, and Kralick and rbis by Rollins and Allison (2) to distribute between the third and fifth innings. We also would like to know if Allison homered in the third or fifth inning.

Could Allison have scored in the third? If so, it was in his second plate appearance. Then since there must be at least 7 batters in the third and fourth innings, Kralick would have had his second (and last) plate appearance before the fifth inning, so he couldn't have scored any runs. So Allison must have scored (and homered) in the fifth.

Who scored the other run of the fifth? It can't be Kralick because the four players between Kralick and Allison would have to be out for Allison to score. So it was Power who scored the other run, and Kralick who scored in the third inning.

Solution to In Scoring Position (p. 94)

TWINS 11, INDIANS 10

Cleveland	AB	R	H	BI	BB	SO	Avg
Lofton cf	4	1	2	0	0	0	.266
b Cabrera ph ss	1	1	1	0	0	0	.333
Vizquel ss	4	1	1	0	0	1	.254
c WCordero ph	0	1	0	0	1	0	.318
RAlomar 2b	4	1	1	4	0	0	.343
JGonzalez rf	4	2	3	2	1	0	.359
Thome 1b	4	1	3	3	1	1	.285
Burks dh	4	1	1	1	1	1	.309
MCordova lf cf	5	0	0	0	0	0	.353
Fryman 3b	5	1	2	0	0	2	.222
Taubensee c	3	0	0	0	0	1	.242
a EDiaz ph c	2	0	0	0	0	1	.313
Totals	40	10	14	10	4	7	

Minnesota	AB	R	H	BI	BB	SO	Avg
J.Jones lf	6	1	2	0	0	3	.264
CGuzman ss	6	3	4	2	0	1	.296
Mientkiewicz	4	1	1	0	1	1	.344
Lawton rf	5	2	3	0	0	0	.290
THunter cf	5	0	1	0	2	2	.261
Koskie 3b	2	1	1	0	3	1	.245
Buchanan dh	4	1	1	2	1	2	.212
Pierzynski c	4	0	2	2	0	2	.274
d Hocking ph	1	0	0	0	0	0	.269
Rivas 2b	5	2	2	0	0	1	.221
Totals	42	11	16	10	5	13	

E—Vizquel (4), Taubensee (2), Rincon (1), Pierzynski (5). LOB—Cleveland 8, Minnesota 11. 2B—Thome (10), J.Jones (9), Mientkiewicz (15), Lawton 2 (12) Buchanan (2). 3B—Vizquel (2). HR—Burks (13) off TMiller, Thome (14) off Mays, JGonzalez 2 (15) off BWells, Mays, RAlomar (5) off Mays. RBIs—RAlomar 4 (34), JGonzalez 2 (52), Thome 3 (35), Burks (40), CGuzman 2 (16), Lawton 3 (23) THunter (30), Buchanan 2 (11), Pierzynski 2 (15). SB—Lofton (7), Koskie (7), Rivas (9). SF—RAlomar. Runners left in scoring position—Cleveland 4 (Burks, Fryman 3) Minnesota 7 (J.Jones, Lawton, Buchanan 2, Pierzynski 3).

Cleveland	IP	H	R	ER	BB	SO	NP	ERA
Burba	1.2	7	8	6	3	3	54	6.02
Westbrook	4.1	3	0	0	2	6	76	3.86
Rincon	1	3	2	1	0	1	20	4.05
Shuey L	1.1	3	1	1	0	3	26	2.43

Minnesota	IP	H	R	ER	BB	SO	NP	ERA
Mays	.6	9	5	5	2	4	108	2.98
BWells	0.1	1	1	1	0	0	5	4.05
TMiller	0.1	1	1	1	0	1	11	5.00
Cresand	0.1	1	1	1	0	0	9	3.09
Guardado W	2	2	2	2	2	2	30	4.85

Mays pitched to 1 batter in the 7th, Cresend pitched to 1 batter in the 8th. Inherited runner scored—Westbrook 2 2, BWells 1 1, Guardado 1 1. IBB—off Guardado (JGonzales), 1 off Westbrook (Koskie) 1. Umpires—Wieters, Barrell, Ted, Marquez, Rippley. T—3:36. A—20,617.

Indians . 2 0 1   0 1 0   3 3 0—10 14 3
Twins . . 2 6 0   0 0 0   2 0 1—11 16 1

One out when winning run scored. a struck out for Taubensee in the 8th. b singled for Lofton in the 8th. c walked for Vizquel in the 8th. d grounded out for Pierzynski in the 9th

We start by looking at who can drive in whom.

slot		runs	rbis	can drive in
1a	Lofton	1		
1b	Cabrera	1		
2a	Vizquel	1		
2b	Cordero	1		
3	Alomar	1	4	1a,1b,2a,2b,3,8
4	Gonzales	2	2	1a,1b,2a,2b,3,4,8
5	Thome	1	3	1a,1b,2a,2b,3,5
6	Burks	1	1	1a,1b,2a,2b,3,4,5,6
7	Cordova			
8	Fryman	1		
	Taubensee			
9	Diaz			

Alomar and Thome, each with a homer and only 1 run, clearly drive in their own runs. Gonzales has 2 homers and just 2 rbis, so he hits two solo home runs. That leaves Alomar as the only guy who can drive in Fryman.

slot		runs	rbis	drives in	and two of
3	Alomar	1	4	3,8	1a,1b,2a,2b
4	Gonzales	2	2	4,4	
5	Thome	1	3	5	1a,1b,2a,2b
6	Burks	1	1	6	

PA = 27 + 10 + 8 = 45. Every position has 5 plate appearances so the players who are walked are Vizquel, Alomar, Gonzales (intentional walk), Thome, and Burks. The last batter of the game will be Diaz in the ninth slot.

We're going to start with the beginning of the eighth inning and slowly work backwards in time, using the box score abundantly. There are 4 pitchers in the seventh inning, in order: Mays (1 batter, no outs), then Wells (HR to Gonzales, 1 out), then Miller (HR to Burks, 1 out), then Cressend (1 out in seventh, 1 batter in eighth). This will help us map out the seventh.

We start by noting that Diaz strikes out in the eighth. Cressend pitches to just 1 batter in the eighth and has no strikeouts, so he must pitch to Fryman or someone earlier in the lineup. So the last batter of the seventh inning must be Cordova or someone still earlier.

Before Cordova is Burks, who homers off Miller. That leaves Cordova as the only batter Cressend faces in the seventh. He is retired by Cressend.

Before Burks is Thome and before him is Gonzales, who homers off Wells. Miller gets one man out. That man must therefore be Thome.

Gonzales' homer is a solo homer—he can't drive anybody else in. So Alomar before him must score or be out before Gonzales homers. If he scores, it must be on his homer, since he has only one run.

Alomar's homer is off Mays. If he homered, Mays pitched to him, and Wells only pitched to 1 batter, Gonzales. This is impossible, because Wells is credited with getting an out, which he couldn't do because the bases would be empty when he came in to pitch to Gonzales. So Alomar didn't homer, and made an out, charged to Wells.

We've accounted for all 3 outs of the seventh—Alomar, Thome, and Cordova—so any player before Alomar in the inning must score. We need Vizquel to bat and score to have 3 runs—Vizquel, Gonzales, and Burks. Vizquel must therefore be the first batter in the inning, put on base by Mays. No one but Alomar can drive him in. We already proved Alomar didn't homer here, but he could have driven him in with either his sacrifice fly or a ground out.

We return now to the eighth and move forward in time. In the eighth, Fryman leads off. He must get on base, since Cressend is charged with a run and pitches only to Fryman in this inning. Cordova is hitless so Fryman must get the one hit that Cressend gives up, a single.

Cressend is then replaced by Guardado. Guardado strikes out Diaz in his first at-bat. Now the pinch hitters, Cabrera and Cordero bat. Cabrera singles and Cordero walks. By the Gap Rule, the three who score this inning must be among the first 5 batters. Alomar doesn't score (he scores his one run when he homers off Mays). So Fryman, Cabrera, and Cordero score and they load the bases when Alomar comes to the plate.

Fryman, as we have proved, is driven in by Alomar. Alomar's only hit is his home run off Mays which must be earlier. He could drive in Fryman with his sac fly, with a ground out. A bases-loaded walk is not possible since Guardado gives up only 2 walks, 1 to Cordero and 1 to Gonzales.

Who drives in Cabrera and Cordero? Guardado intentionally walks Gonzales (see the box score). The only players who can drive in Cabrero and Cordero are Alomar and Thome (Burks and Gonzales drive in no one but themselves; the inning ends with Burks). In addition to Fryman, Alomar can drive in at most Cabrero (he drives in Vizquel and also himself earlier in the game) so Thome drives in Cordero and we can't be sure who, Alomar or Thome, drives in Cabrero.

The inning ends here with Burks since Cordova, Fryman, and Diaz will be the 3 batters in the ninth. If Thome gets to second base, then this is the inning that Burks leaves a runner in scoring position.

Turning to the issue of RLSP, note that Fryman bats in the sixth, eighth, and ninth but doesn't end any of those innings. He must therefore, in his 2 earlier plate appearances, end innings with runners (a total of three) in scoring position. Lofton ends the sixth inning, so Fryman bats in the fifth and can't end the fourth. Fryman can't end consecutive innings since that would require an inning to have 9 PA and 3 runs. Thus, Fryman must end the first and third innings.

The third and fifth innings must have at least 5 PA (both have a run and a man left in scoring position).

When Fryman comes up in the first, seven men have batted before him; two made outs; two scored, so the bases are loaded. When Fryman ends the inning he will have left two on base in scoring position. In the third he will leave another for a total of three. From the above scorecard, it's clear that Burks can't end any inning except the eighth. Therefore Thome did indeed reach second in the eighth and was left in scoring position by Burks.

Lofton must score in the first, driven in by Alomar. The three homers, one each by Alomar, Gonzales, and Thome are in the first, third, and fifth innings. We can't rigorously prove when they occur except to say that Thome's homer can't be in the first because three would have to be out before he could hit a solo homer. The homers are clearly not in the order mentioned in the box score, since Burks' homer is not the first.

Cleveland	1	2	3	4	5	6	7	8	9	10	11	ab	R	H	rbi
Lofton/Cabrera												5	2	3	0
Viquel/Cordero												4	2	1	0
Alomar												4	1	1	4
Gonzales							HR	IBB				4	2	3	2
Thome												4	1	3	3
Burks							HR	RESF				4	1	1	1
Cordova												5	0	0	0
Fryman	RESF		RESF									5	1	2	0
Taub./Diaz							K					5	0	0	0
TOTALS	2	2	1	1	1	1	3	3				40	10	14	10

Solution to Just Totals (p. 97)

Our two equations for plate appearances (p.9 and p.34) give us:

$$R + LOB + Outs = AB + BB + HBP + S + SF,$$

which gives us an equation for the number of outs in the game:

$$Outs = AB + BB + HBP + S + SF - R - LOB.$$

In this case,

$$Outs = 93 + 8 + 0 + 0 + 0 - 5 - 18 = 78.$$

$78 = 13 \times 6$. That means the teams played a full 13 innings. The visiting team, Chicago must have won. If Cleveland had won it would have been a walk-off, with Cleveland having fewer than 3 outs.

Note that if AB had been 92 instead of 93, we would have concluded that Cleveland won in a walk-off with 2 outs in the 12th inning.

The game, by the way, was played on September 30, 2005.

Bonus Puzzle!

As we write this, major league rules are changing. One change, initially made to shorten games during the pandemic, seems likely to stay—the "ghost runner." The rule is that every half inning beyond the ninth begins with a player on second base. The player placed on second is the player who batted last in the previous inning. For the records, this isn't considered an at-bat for the player. No other details are needed for the following puzzle:

The Cubs played the Yankees in New York (real game, 2022). Here are the non-zero totals for the two teams together:

AB	R	H	BI	BB	SO	DP	2B	HR	LOB	SB	CS	HBP
89	3	15	3	12	25	3	3	2	30	1	1	1

Who won?

Solution on p. 234

Solution to Two Guys (p. 98); (Hint on p. 173)

Hall scored 4 runs, all driven in by Dominguez. Each time, Hall must lead off, because otherwise Dominguez would be the 7th player to bat in the inning. Of the previous six players, only three can be on base, only two can be out, and none can have scored (because only Hall scores in the game and then only when driven in by Dominguez).[3]

Hall, then, has to be the first one up four times in the game. The first

Cats

	ab	r	h	bi
Abe ss	4	0	1	0
Brown 2b	4	0	1	0
Castro rf	4	0	1	0
Dominguez lf	3	0	2	4
Engel c	5	0	0	0
Frank cf	4	0	1	0
Guzman 3b	4	0	0	0
Hall 1b	2	4	2	0
Iglesias p	3	0	1	0
Totals	33	4	10	4

possible inning is the third (you can't have seven up in the first inning without a run scoring). And Hall can't be first up in 2 consecutive innings, so if the game has only 9 innings, Hall has to lead off in the third, fifth, seventh, and ninth. And indeed the game must have only 9

---

[3]This is like the situation in A World Series Game, p.197.

innings—in extra innings, the Cats would have to score 2 runs in 1 inning to win 4-2, which impossible with only Hall scoring.

So in the third, fifth, seventh, and ninth innings Hall scores. In each of those innings, Dominguez comes to the plate with the bases loaded, 2 out, and Hall on third. In the third, fifth, and seventh, Dominguez must be the last player to bat in those innings so that Engel, Frank, and Guzman can bat in the next inning (so Hall can lead off in the inning after that). Note that this is possible (one way, for example, is if Dominguez hits a single and the player who was on second makes the third out trying to score on the play).

Thus, Engel leads off for the Cats in the sixth, answering one question. The Cats leave two on base in the seventh (with 1 run and 3 outs, it takes 2 LOB to make 6 PA), answering the second question. The critical contributions of Hall and Dominguez make the two of them the natural answer to the "two guys" of the third question.

Solution to Smedley (p. 99)

Smedley had the last PA of the game, and thus the last PA for his team. He was a starter, say in slot x. His third and final PA was the x+18th of the game, so his team had at most 9+18 or 27 PA.

Since Smedley's team had 2 LOB, it can have at most 25 outs, already making the maximum 27 PA. A 9-inning game can't be completed if the visiting team has less than 27 outs, so Smedley's team must have been the home team and won. To win it must have scored at least one run. It couldn't be more than 1 run, since 1 run + 2 LOB + 24 outs in the first 8 innings is already the maximum possible 27 PA. So Smedley's team won 1-0.

Since Smedley's final strikeout was the last at-bat of the game it had to be in the ninth, after the visiting team's half of the ninth. Smedley's team can't have scored before the ninth inning, or Smedley wouldn't be able to end the game—the game would end with the visitor's ninth inning. Since there are 27 PA in the game, Smedley must bat in slot 9.

There were no outs in the home team's ninth, so Smedley couldn't have been retired when he struck out. The catcher must have failed to catch the third strike (a passed ball or a wild pitch) and somehow Smedley made it to first. Smedley's team must have scored its 1 run on this play, ending the game. Here's one way that could have happened: In the first 8 innings, every player was retired; no one was left on base. The 7 slot hitter led off the ninth and got on base. The 8 slot hitter also got on base. The 7 slot guy scored on the strikeout, leaving two men on base (Smedley and 8 slot). That's it. One run, two left on base, one of them a smirking Smedley on first.

Second Bonus Puzzle!

It's just like the problem above except that instead of ending every inning he's in with a strikeout, Smedley *leads off* every inning he's in with a strikeout. As before, there are two left on base. As before, Smedley is up three times. As before, his last plate appearance, a strikeout, is the last of the game.

The question is the same: What can you say about this?

Solution on p. 237

Solution to Everybody Scores a Run (p. 99)

With 27 LOB in 9 innings there are 3 LOB per inning—the most you can have. There is 1 run in each inning. Hence, there are 7 plate appearances in each inning—unless there are fewer in the ninth.

Blanco drove in six. No one scored more than one. So by the Five-step Rule, Blanco drove in the batters in slots 1, 2, 6, 7, 8, and 9. Furthermore, if Blanco comes up seventh in an inning, he can't drive anyone in—one of the previous six must already have scored.

Dogs

	ab	r	h	bi
Amsler	6	1	1	0
Blanco	7	1	6	6
Clark	6	1	5	0
Durand	3	1	2	1
Emerson	5	1	2	1
Foster	5	1	2	1
Gato	5	1	1	0
Herrera	6	1	1	0
Ito	5	1	1	0
Totals	48	9	21	9

Dogs 1 1 1 1 1 1 1 1 1--9

LOB--Dogs 27, HR--Blanco, 3B--Amsler, 2B--Clark, Durand, Emerson, SF--Blanco, S--Herrera,

All that gives us this chart:

inning	leading off	possible batters driven in by Blanco
1	Amsler, slot 1	1,2
2	Herrera, slot 8	8,9,1,2
3	Foster, slot 6	6,7,8,9,1,2
4	Durand, slot 4	none—Blanco isn't up
5	Blanco, slot 2	2
6	Ito, slot 9	9,1,2
7	Gato, slot 7	7,8,9,1,2
8	Emerson, slot 5	none—Blanco is 7th
9	Clark, slot 3	none—Blanco isn't up

Blanco can't drive in anyone in innings 4, 8, and 9. So he must have driven in runners in all the other innings. So he must have driven in himself in the fifth—that's the only possibility in that inning. That's his solo home run.

Then in the first he must have driven in Amsler. And we keep going by elimination: Blanco drives in Ito in the sixth, Herrera in the second, Gato in the seventh, and Foster in the third.

Who drove in Clark, Durand, and Emerson in the fourth, eighth, and ninth innings? In the eighth, the only one of those three who could be driven in is Emerson and the only one who could do it is Foster.

Then in the fourth, that leaves Durand who must have been driven in by Emerson. And then in the ninth, Clark must have been driven in by Durand. Amsler's triple must be in the first inning, because in any other inning it would have to drive in a run.

That's it.

The following scorecard displays the location of all the runs and rbis as deduced above. It demonstrates that a game consistent with the box score is indeed possible, but we have not proved that the doubles, singles, and walks must be exactly where they are placed.

DOGS	1	2	3	4	5	6	7	8	9	10	11	ab	R	H	rbi
Amsler	(3)	•	•	•		•	•	w				6	1	1	0
Blanco	1	1	1		HR	1	1	•				7	1	6	6
Clark	1	1	•		1	1	w	(2)				6	1	5	0
Durand	w	w		(2)	w	w	•		1			3	1	2	1
Emerson	•	•	1		w	•	(2)	w				5	1	2	1
Foster	•		(1)	w	•	•		1	w			5	1	2	1
Gato	•		w	w	•	(1)	•	•				5	1	1	0
Herrera	(1)	w	•	•			•	•	•			6	1	1	0
Ito		•	•	•		(1)	w	w	•			5	1	1	0
TOTALS R H	3	3	2	2	2	3	2	2	2			48	9	21	9

Solution to Bonus Puzzle (p. 230)

We'll start with the calculation in the solution to Just Totals:

$$\text{Outs} = 89 + 12 + 1 + 0 + 0 - 3 - 30 = 69.$$

But this isn't correct for games with extra innings and the new ghost runner rule. The ghost runner doesn't have a plate appearance, but his placement on second base has the same effect as one. Each ghost runner either makes an out, scores a run, or is left on base—just like players who had a plate appearance. It turns out that in our calculation of outs,

$$\text{Outs} = \text{AB} + \text{BB} + \text{HBP} + \text{S} + \text{SF} - \text{R} - \text{LOB}.$$

we have to add $2X$, where X is the number of extra innings.

$$\text{Outs} = 2X + \text{AB} + \text{BB} + \text{HBP} + \text{S} + \text{SF} - \text{R} - \text{LOB}.$$

If we subtract 54 from this (the number of outs in the first nine innings) we have the total extra-inning outs. In the current puzzle, that's

$$\text{Extra-inning outs} = 2X + 69 - 54 = 2X + 15$$

In a walk-off, the last extra inning will have fewer outs. Let q be the number of fewer outs (if the visitors win q will be zero).

But this should equal $6X - q$, 6 outs in X extra innings minus q. From

$$2X + 15 = 6X - q$$

we have

$$15 = 4X - q$$

From this, X must be 4 and $q = 1$. Thus, the Yankees defeated the Cubs with a 2-out walk-off in the 13th inning.

Note that a simple way to find out who won is to just to ask, is

$$AB + BB + HBP + S + SF - R - LOB - 54$$

a multiple of 4? If it is, the visitors won. If it isn't, the home team won.

The game, by the way, was played on June 10, 2020.

Solution to <u>Castro Homers in the First</u> (p. 100); (Hint on p. 173)

We definitely have that in every inning 4 men come to the plate, except maybe the ninth. Leading off the innings, then, in order, are 1, 5, 9, 4, 8, 3, 7, 2, and 6, a different guy each time.

Each player who hits a HR drives in himself, so he can't drive in any other player. So we have 6 players driving in themselves, and 3 players driving in each other. Call these players x, y, and z. We can't have x driving in y and y driving in x, because then z would have to drive in himself, and that would require a 7th HR. So we must have a cycle: x drives in y, y drives in z, and z drives in x. Or the other way around. Call these guys 'cyclists'.

How would this cycle work? There are at most 4 batters up in an inning, so there can be at most 2 players between the cyclist who scores and the one who drives him in. Here are three cycles that can work: $_4\!\!\overset{1}{\leftarrow}\!_7$ (4 driving in 1, 7 driving in 4, and 1 driving in 7), $_5\!\!\overset{2}{\leftarrow}\!_8$, and $_6\!\!\overset{3}{\leftarrow}\!_9$.

But we're given that the third batter hits a homer in the first inning. Thus the cycle can't be $_4\!\!\overset{1}{\leftarrow}\!_7$ (because 4 can only drive in 1 in the first inning, when 1 leads off and 3's homer would drive in 1 before 4 came to bat). It also can't be $_6\!\!\overset{3}{\leftarrow}\!_9$ (because Castro hit a homer) so it has to be $_5\!\!\overset{2}{\leftarrow}\!_8$.

Cats	1	2	3	4	5	6	7	8	9	10	11	ab	R	H	rbi
Abe	•		•		•		•						1	1	1
Brown	•		•	•			◇						1	1	1
Castro	HR		•			•		•					1	1	1
Dominguez	•			•		•		•					1	1	1
Engel		◇		•		•		•					1	1	1
Frank		•		•		•			•				1	1	1
Guzman		•		•			•	?					1	1	1
Hall		•			◇		•	?					1	1	1
Iglesias			•		•		•	?					1	1	1
TOTALS													9	9	9

From this we can see that in the sixth inning, the homer isn't the third player's or the fifth player's. It also can't be the sixth player, since it can't be the last batter in an inning, so it must be the fourth player's.

Look next to the fourth inning and you can see it must be the sixth player's homer. In a similar way you will find there is only one way to arrange the homers.

Cats	1	2	3	4	5	6	7	8	9	10	11	ab	R	H	rbi
Abe	•	HR		•		•						4	1	1	1
Brown	•		•	•			◉					3	1	1	1
Castro	HR		•			•		•				4	1	1	1
Dominguez	•			•	HR			•				4	1	1	1
Engel		◉		•		•		•				3	1	1	1
Frank		•	HR	•					•			4	1	1	1
Guzman		•		•			•	HR				4	1	1	1
Hall		•			◉		•					2	1	1	1
Iglesias			•		•	HR						3	1	1	1
TOTALS												31	9	9	9

Note that Hall, Brown, and Engel were all tagged out after driving in their runs.

Finally, since every inning had one hit, the homer by #7 is the only hit in the ninth inning. We're told it ended the game, so we know that Cats won in a walk-off. Note that the ninth inning doesn't have to have 4 at-bats, since it doesn't have to have 3 outs.

Solution to Brown Homers in the First (p. 101)

Following the reasoning in the solution to the previous puzzle we get a different picture,

Cats	1	2	3	4	5	6	7	8	9	10	11	ab	R	H	rbi
Abe	•		•		HR		•					4	1	1	1
Brown	HR		•		•			•				4	1	1	1
Castro	•		•			•		•				3	1	1	1
Dominguez	•			•		•		HR				4	1	1	1
Engel		•	HR		•			•				4	1	1	1
Frank		•		•		•		•				3	1	1	1
Guzman	HR		•			•		•				4	1	1	1
Hall		•		•	HR		•					4	1	1	1
Iglesias		•		•		•		•				3	1	1	1
TOTALS	R/H											33	9	9	9

and a different winner. The pitcher drove in a run in the last at-bat. But there are none left on base for Cats so the pitcher must get out somehow before the next at-bat is completed. If Cats had won, the game would have ended before that out. Hence the Dogs won.

Solution to Second Bonus Puzzle (p. 232)

Smedley is the first and last batter in the ninth, hence the only batter. He strikes out. How is this possible? One out can't end the inning. Instead, Smedley must somehow score a walk-off run. With 24 outs, 2 LOB, and Smedley's run there will already be 27 PA in the game, the maximum possible for Smedley, the final batter, to have only 3 PA.

Smedley would have to bat in slot 9, since his third PA is the team's 27th. For Smedley to lead off the third inning, the 2 LOB must occur in the first 2 innings. So Smedley leads off the third, the sixth, and the ninth, and before the end, no run has been scored. What must happen in the ninth is this: Smedley strikes out. The catcher fails to catch the third strike. Smedley runs to first. Then in a series of errors, he races all around the bases and scores. Or maybe Smedley, once he gets on base, steals second, third, and home. In either case, the game is over. Nobody is out. Nobody is left on base.

And Smedley is the hero.

Solution to <u>Leftovers</u> (p. 102); (Hint on p. 174)

	ab	r	h	bi
Alonso ph, 2b	2	1	2	0
Chen rf	1	1	1	0
Croteau ph, rf	3	0	0	0
Faatz p	1	0	0	0
Gil ph, lf	1	0	1	1
Hanlon cf	3	1	2	1
Jackson p	0	0	0	0
Kelly lf	3	0	3	2
Kimura p	2	0	0	0
Maeng 2b	2	0	1	1
Pasig 3b	4	2	2	0
Tabeau 1b	3	0	1	1
Yoshida c	4	1	0	0
Zhao ss	3	0	1	0

	ip	h	r	er	bb	so
Martin (W)	9	14	6	6	0	4

LOB—Eastham 9, Westwood 5, HR—Croteau, Namath, Okada, DP—Westwood 6, 3B—Hanlon (2), HBP—Duffy by Faatz, GIDP—Pasig, Yoshida (3), Hanlon, Faatz, SF—Hanlon

Westwood   4 0 0 0 0 0 0 0 2—6

Westwood has 6 runs, 27 outs, and no LOB for a total of 33 PA. The last slot up is slot 6. There are, thus, 6 slots with 4 PA and 3 with 3 PA. With no LOB we can locate all the plate appearances in the scorecard. The first inning will have 7 slots batting. The last inning will have slots 2-6 batting.

	1	2	3	4	5	6	7	8	9
	•	•			•			•	
	•		•			•			•
	•		•			•			•
	•		•			•			•
	•			•			•		•
	•			•			•		•
	•			•			•		
		•			•			•	
		•			•			•	
R/H	4								2

The key innings are the first with 4 runs and the last with two. We see immediately that Pasig, Yoshida, and Hanlon are in the first six by their 4 plate appearances (Hanlon has 3 AB plus the SF).

From the chart of plate appearances and the fact that all runs are in the first and ninth innings, the first 6 slots must include all those who score runs: Alonso, Chen, Hanlon, Pasig, and Yoshida. The first 7 slots must in addition include all those who drive in runs: Gil, Kelly, Maeng, and Tabeau.

Together these players have a total of 27 PA. That's exactly as many PA as there are in slots 1-7, so the remaining players, Zhao, Kimura, and Faatz, fill slots 8 and 9. Clearly that means Zhao takes one slot and the pitchers take the other. Faatz, who enters the game late and hits into a double play, can't be in slot 8, which leads off the eighth inning. Thus Zhao is in slot 8 and Kimura/Faatz is in slot 9.

We know from Scraps that the pitcher Faatz follows Kimura. Pitcher Jackson is never in the lineup as he joins the team and leaves it entirely while his team is on the field in the seventh inning.

There are substitutions. Croteau must substitute for a player with 1 AB. That can't be Gil (who substitutes for someone else) and it can't be Faatz who doesn't enter the game until after Kimura has had 2 at-bats. Thus it's Chen, and Chen/Croteau has 4 PA and one of the first 6 slots.

There are four substitution pairs based on fielding positions: Maeng/Alonso, Chen/Croteau, Gil/Kelly, and Faatz/Kimura. There could be a double switch, if two substitutions happen at the same time. But Faatz and Kimura are already a slot pair in slot 9. Maeng and Alonso must be slot-paired as they are the only 2 with 2 PA each. In theory there could be a double switch involving the two 3PA-1PA pairs, except that Croteau (3PA) and Gil (1PA) are not starters and who substitute at clearly different times, so Chen/Croteau share a slot and Kelly/Gil share a slot.

The runs are in the first and ninth innings. The player in slot 7 can't score in the first (being last up) or in the ninth (not having an at-bat) but can have rbis in the first. Possibilities are Kelly/Gil and Tabeau. But Kelly/Gil has rbis in both the first and the ninth, so it must be Tabeau in slot 7.

The 2 runs in the ninth are scored by Alonso, who is not in the lineup in the first inning, and Pasig, who scores in both innings. Thus, Chen, Hanlon, Pasig, and Yoshida score in the first while Maeng, Kelly, and Faatz are retired. The rbis in the ninth are by Gil (not in the lineup in the first) and Hanlon (his sacrifice fly—Hanlon can't have both the SF and the run in the first, that would require a fielding error on Hanlon's fly ball and there are no errors in the game by either team). All the other rbis (Maeng, Kelly (2), and Tabeau) are in the first.

We need to place Kelly/Gil, Pasig, Chen/Croteau, Yoshida, Hanlon, and Maeng/Alonso in the first six slots.

Maeng/Alonso and Pasig score in the ninth so they can't be in slot 1. Neither can Yoshida, who, without a hit couldn't get on base leading off the first. Kelly/Gil can't be in slot 1 as Gil drives in a run in the ninth. Hanlon also drives in a run in the ninth. That leaves Chen/Croteau in slot 1.

Pasig, Yoshida, and Hanlon can't be in slot 2 because they ground into double plays (impossible in the first, since Chen scores) and impossible later since slot 2 leads off subsequent innings. Kelly can't be in slot 2 because Kelly drives in two in the first and doesn't score. That leaves Maeng/Alonso in slot 2. Maeng drives in Chen.

Pasig has a run in the ninth, so he can't be in slot 6. If he were in slot 5 he wouldn't be able to ground into a double play (slot 5 leads off the fourth and the seventh and Pasig scores in the first and the ninth). Thus, he must be in either slot 3 or 4.

We're going to need, in a moment, that neither Yoshida nor Hanlon grounds into a double play in the first inning. While it's possible to ground into a double play without being retired, you need two men on base for that. In the first inning that would be two of Maeng, Kelly, and Tabeau (the three who don't score). Since Tabeau is in slot 7, that would mean that Yoshida or Hanlon must follow Maeng and Kelly and precede Pasig (who can't score before Maeng and Kelly are retired). But that's impossible since we know Pasig is in either slot 3 or 4, so there's no way Yoshida or Hanlon could ground into a double play in the first.

Since Yoshida doesn't GIDP in the first, he GIDPs in the ninth, so Hanlon can't be in slot 6 because his SF (which we showed must be in the ninth) must precede Yoshida's GIDP (there can be only one out to have a sacrifice fly). Hanlon's own GIDP means he can't be in slot 5. Consequently Pasig and Hanlon are in slots 3 and 4. If Pasig is in slot 3, he will be driven in by Hanlon's triple in the first, but Hanlon has no rbi in the first. Hence Hanlon is in slot 3 and Pasig is in slot 4.

Yoshida can't be in slot 5 (those GIDPs) so he's in slot 6. That leaves Kelly/Gil in slot 5. That's the lineup.

Westwood	1	2	3	4	5	6	7	8	9	10	11	ab	R	H	rbi
Chen/Croteau	①	•		•			•					4	1	1	0
Maeng/Alonso	•		•			•			①			4	1	3	1
Hanlon	③		3		DP			SF				3	1	2	1
Pasig	①		DP			•			①			4	2	2	0
Kelly/Gil	•			•			•		•			4	0	4	3
Yoshida	•			DP			DP	DP				4	1	0	0
Tabeau	•			•			•					3	0	1	1
Zhao		•			•			•				3	0	1	0
Kimura/Faatz		•			•			DP				3	0	0	0
R/H	4/7		/1						2/3			32	6	14	6

Additional notes to explain how this scorecard could have occurred: Kelly and Tabeau could be picked off base shortly after their at-bats in the first, or they could have been thrown out trying to take an extra base. Pasig's single in the first could be a bunt fielded by the third baseman who didn't attempt a throw to first but kept the ball to hold the runner on third. The double play in the third might have been a short-hop grounder to third enabling the third baseman to tag Hanlon off the base and then throw to first.

Solution to Four Players Missing (p. 105)

The player in slot 3 (call him "#3") can't score in the second inning because the first inning can't have more than 7 plate appearances, so the earliest #3 can bat in the second inning is as the fifth batter. So #3 scores in the first. He can't be driven in by #7, because if #7 comes to the plate in the first, at that point

	ab	r	bi
1.	4	0	0
3.	3	1	0
5.	4	0	0
7.	3	0	1
9.	3	0	0
	30	2	2

1 1 0 0 0 0 0 0 0—2

there can be at most 2 outs and 3 on base, so the run by #3 must have already been scored. So #3 is driven in by #4 or #6. In either case, #4 must bat in the first inning.

So #7 must have his rbi in the second inning, and since #5 didn't score any runs and #4 didn't come to bat in the second, #7 must have driven in #6. That means #6 didn't bat in the first inning, so only #4 could have driven in #3 then.

Solution to Four Missing, Again (p. 106)

All the runs (and none of the rbis) were scored by #s 2, 4, 6, and 8. The sole run in the first inning has to be #2, and must be driven in by #3.

To score 3 runs in the fourth inning, the scorers have to be #s 2, 4, 6, with #2 leading off, or #s 4, 6, 8, with #4 leading

	ab	r	bi
1.	4	0	1
3.	3	0	2
5.	4	0	0
7.	3	0	1
9.	3	0	1
	30	5	5

1 0 0 3 1 0 0 0 0—5

off. Nothing else is possible by the Gap Rule (the first three to score in an inning must be among the first 5 batters—p.24).

Could #s 2, 4, 6 score in the fourth? Let's see. The rbis must be from #3, #7, and #9, since #1 can't drive in #6 (#3 and #5 would have to be out and also one of #7, #8, #9, making 3 outs). But then #1 will have to drive in someone in the fifth and that's not possible since either #1 or #2 will lead off the fifth.

Thus it must be #s 4, 6, and 8 who score in the fourth. #4 leads off the fourth inning. #5 makes an out. #6 gets on. #7 drives in #4. #8 gets on. #9 drives in #6. Somehow #7 makes the second out. Then #1 drives in #8. #1 makes it safely to first, but either #1 or #9 makes the third out.

Fifth inning:

#2 leads off and is again driven in by #3.

	ab	r	bi
1.	4	0	1
2.		2	0
3.	3	0	2
4.		1	0
5.	4	0	0
6.		1	0
7.	3	0	1
8.		1	0
9.	3	0	1
Totals	30	5	5

1 0 0 3 1 0 0 0 0--5

Solution to Who Won? (p. 114); (Hint on p. 175)

Cats	ab	r	h	bi
Abe				
Brown				
Castro				
Dominguez				
Engel				
Frank				
Guzman				
Hall				
Iglesias p	1	0	0	0
Ivory ph	1	0	1	1
Ingals p	1	0	0	0

	ip	h	r	er	bb	so
Cats						
Iglesias	5	7	3	3	1	2
Ingals	4	3	3	3	1	4
Dogs						
Ito	5	2			0	15
Ignacio	4				3	2

LOB—Cats 2, Dogs 8, DP—Cats 2, 3b—Hall, 2b—Emerson

Since the visiting team is always listed first in the pitching section of a box score, we know the Cats are the visiting team.

How many plate appearances did Iglesias have in innings 1–5?:

Only one plate appearance. Iglesias has 1 at-bat, no S, SF, or HBP. Further, he has no walks in those innings since Ito, who pitched the first 5 innings, gave up no walks.

Which batting slot leads off the sixth inning?:

Ito pitches 5 innings, giving up 2 hits and striking out 15 batters, accounting for at least 17 plate appearances. He can't have pitched to any more, because the 18th would have been a second PA for Iglesias (Ivory wouldn't pinch-hit for him in the first 5 innings because Iglesias, on the visiting team, still has the bottom half of the fifth to pitch). So Ito pitched to exactly 17 batters and the 9 slot led off the top of the sixth.

When did Ivory drive in a run?:

Ivory must pinch hit for Iglesias while Iglesias is still in the lineup and Ingals has not yet pitched, namely in the top of the sixth inning. He can't bat in any later inning, because he is replaced by Ingals in the bottom of the sixth.

Ivory's batting slot starts the sixth. Ivory has an rbi but no homer, so even if he leads off the sixth, he can't get a rbi as a leadoff batter. Thus, the Cats must have 10 batters up in the sixth for the 9 spot to come up again so Ivory can get an rbi.

That gets us to 27 plate appearances in 6 innings; hence at least 36 for 9 innings. With 2 LOB and 27 outs Cats must have scored at least 7 runs. So they won the game, 7-6.

Here's what a scorecard might look like:

Cats	1	2	3	4	5	6	7	8	9	10	ab	R	H	rbi
Abe	K		K			1 •					4	1	1	0
Brown	K			K		W					3	1	0	0
Castro	K			1		1 •					4	1	2	2
Dominguez		1		K		K	•				4	0	1	0
Engel		K		K		K	•				4	0	0	0
Frank		K			K	W	•				3	1	0	0
Guzman		K			K	1		•			4	1	1	1
Hall			K		K	3		•			4	1	1	3
Iglesias			K								1	0	0	0
Ivory						W 1-					1	1	1	1
Ingals								•			1	0	0	0
TOTALS	R H		1	1		7 5					33	7	7	7

Solution to Sad Story (p. 115)

Someone scores in the first. That has to be one of the first 3 batters, so it must be Amsler. Every run is batted in. Emerson has only one rbi and that must be himself on his homer, so the only one who can drive in the #1 batter is Foster. This happens with 2 out and the bases loaded with Emerson on first base. Emerson must have singled and there is no way he can steal a base this inning.

The other 3 runs for the Dogs are a solo homer by Emerson, a run by Durand driven in by Gato, and a run by Emerson also driven in by Gato.

Skipping to the eighth inning, where 2 runs are scored, Emerson can't homer in that inning because then Gato would be unable to drive in either him or Durand, so there would be no way to score a second run. That means that Emerson homers in the third inning and in the

Dogs batting

	ab	r	h	bi
Amsler		1		
Blanco				
Clark				
Durand		1		
Emerson	4	2	4	1
Foster				1
Gato				2
Herrera				
Ito				
Totals				

Cats	·· 0 0 0	1 2 0	0 1 1--5
Dogs	·· 1 0 1	0 0 0	0 2 0--4

Cats Pitching

	ip	h	r	er	bb	so
Iglesias (W)	8		4	4	0	6
Ivory	1	0	0	0	1	2

LOB- Dogs 10, HR-Emerson, 3b-Emerson, 2b-Emerson, Gato, Amsler, SB-Emerson (2)

eighth Gato drives in both Durand and Emerson. Emerson can't walk in the eighth, because Iglesias, who pitched the eighth, gave up no walks. Emerson can't triple in the eighth because then Durand wouldn't still be on base to be driven in by Gato. So Emerson doubles in the eighth, moving Durand to third. Durand has to stay on third until he is driven in by Gato, so Emerson can't steal in this inning either.

Emerson had 5 plate appearances. By our equation, the Cats had $27 + 10 + 4 = 41$ PAs so Emerson's fifth plate appearance was the last of the game. He didn't get a hit because Ignacio, who pitched the ninth, gave up no hits. Thus, Emerson's triple came earlier. We can't say in which inning, but we can say he didn't steal in that inning either, because he would have had to steal home for a second run, when he had only one.

Thus, Emerson stole both his bases in the ninth. He only has 4 at-bats, so he must have Ignacio's walk. He then stole second and third. There is no further plate appearance. Nobody else could be on base after he stole second and third. He must have been thrown out on the bases for the third out, either picked off base or thrown out trying to steal home. That might be why he felt he was a failure![4]

Solution to <u>Kids' Stuff</u> (p, 117) (hints on p. 175 and p. 177)

Here's one way we can figure things out, in steps: (you can fill out this information in your scorecard as you go along, to see how it works)

Puppies	1	2	3	4	5	6	ab	R	H	rbi
Aiko	?	?	?	K		?	2	1	0	0
Butch	?	?	?		?	?	2	0	0	2
Chulo	?	?	K		?	?	2	1	0	1
Dusty	?	?		?	?	?	2	1	0	0
Em	?	K		?	?	?	2	0	0	1
Fei Fei	?		?	?	?	K	2	1	0	1
Gordito	K		?	?	?		2	0	0	1
Hank		?	?	?	K		2	1	0	0
Izzie		?	?	?		?	2	1	0	0
TOTALS	R H	1 0	1 0	1 0	1 0	1 0	1 0	18	6	0	6

1. Butch has 2 rbis, which can't be in innings 1, 4, 5, or 6, so he must walk and get rbis in innings 2 and 3.
2. The only way Butch can get an rbi in inning 2 is if all 3 batters before him walk, so this must be the case.
3. Only 1 run scores in the second, so Chulo and Dusty must strike out in that inning.
4. For Chulo to have exactly 2 ab, he must walk in his other 3 PAs, in innings 1, 5, and 6.
5. We can now see that Chulo didn't get an rbi in the third, and the only inning left in which an rbi is possible is the sixth, so that's when

---

[4]We can actually say exactly why Emerson thought he was failure. We can do this because we made up the story. The truth is he was down on himself because he had to ask the batboy how to get his cellphone out of airplane mode.

he had his rbi. And that means that the batters before him in the sixth all walked.

6. And also all the batters after Chuko in the sixth struck out, or else there would be 2 runs in that inning.

7. We can now see that Butch has walks in 3 innings and must strike out in the others, innings 1 and 5, in order to have 2 ab. Similarly Izzie must strike out in innings 3 and 4.

8. Since Butch struck out to lead off the fifth, it's no longer possible for Em to have an rbi in that inning. The only possibility left for Em's rbi is in the first inning, so Em walked then and got her rbi then, which means that

9. Aiko and Dusty walked in the first (or else Em wouldn't have been on third base to score) and Fei Fei struck out in the first.

10. Since we have already placed 2 strikeouts for Dusty and Em, Dusty and Em must both walk in innings 4 and 5.

11. Fei Fei also has 2 strikeouts placed, so must walk in innings 3, 4, and 5, getting her rbi in inning 5.

12. The fifth inning is now completed except for Gordito, who must strike out to account for 3 outs, and therefore walks in the third and fourth.

13. Hank must strike out in the fourth, and thus must walk in the third.

14. Aiko strikes out in the third and fourth.

Puppies	1	2	3	4	5	6	ab	R	H	rbi
Aiko	w	w	K	K		w	2	1	0	0
Butch	K	w	w		K	w	2	0	0	2
Chulo	w	K	K		w	w	2	1	0	1
Dusty	w	K		w	w	K	2	1	0	0
Em	w	K		w	w	K	2	0	0	1
Fei Fei	K		w	w	w	K	2	1	0	1
Gordito	K		w	w	K		2	0	0	1
Hank	w	w	K	K			2	1	0	0
Izzie		w	K	K		w	2	1	0	0
TOTALS	R 1 / H 0	1 / 0	1 / 0	1 / 0	1 / 0	1 / 0	18	6	0	6

Solution to <u>Kids Will Be Kids</u> (p. 118); (Hint on p. 176)

Puppies	1	2	3	4	5	6	ab	R	H	rbi
Aiko	?	? K	?		?	?	1	4	0	1
Butch	?	?	? K		?	?	2	2	0	1
Chulo	?	?	?	?	?	?	3	3	0	0
Dusty	?	?	?	?	?	?	2	0	0	0
Em	?	?	?	?	?	?	1	1	0	4
Fei Fei	?	?	?	?	?	K	4	2	0	1
Gordito	K	?	?	?	?		1	1	0	4
Hank		? ?	?		? K		2	1	0	2
Izzie		? ?	?	K		?	2	1	0	2
TOTALS	R 1 / H 0	6 / 0	4 / 0	1 / 0	2 / 0	1 / 0	18	15	0	15

A reminder: since all the outs are made by strikeouts, nobody who walks in an inning is later put out. In each inning with n runs, the first n players who walk score the runs, and the last n players to walk are the ones who have the rbis.

1. Aiko strikes out to end the second, accounting for his only ab, so walks in every other PA. He can't score in innings 3 and 4, so he must score in innings 1,2,5, and 6 to account for his 4 runs.

2. Aiko's run in the sixth is the only run of the inning. Izzie can't walk leading off the sixth, or else he'd score too. So he strikes out in the sixth. Now he has two strikeouts so his other PAs are walks.

3. Gordito's strikeout ending the first accounts for his sole ab, so he walks in his other 4 PA. Since he has 4 rbis, each walk drives in a run.

4. Gordito's rbi in the fourth inning drives in the only run of the fourth, so Hank must strike out after him, or else he'd drive in a second run.

5. Hank's 2 AB are now accounted for, so he must walk in every other PA, namely twice in the second and once in the third.

6. Fei Fei has 2 runs. She can't score in innings 1, 4, 5, or 6, so must walk and score in innings 2 and 3.

7. Fei Fei has 4 AB, so she must strike out in innings 1, 4, 5, and 6.

8. Dusty has no runs, so must strike out in the second and third (since in each inning Fei Fei scores later in the inning). This accounts for his 2 AB, so he must walk in his other PAs.

9. Since Aiko scores after Hank's and Izzie's first walks in the second, Hank and Izzie score then too.

10. Since Gordito's walks drive in runs in the second and third, so do all the walks that immediately follow (two in the second, three in the third). In particular, Aiko's walk in the third drives in Gordito.

11. Em has 4 rbis. She can't have one in the third or fourth; hence she walks and drives in a run in innings 1, 2, 5, and 6, and strikes out in either the third or the fourth). But she can't strike out in the fourth (there are already three strikeouts, Fei Fei, Hank, and Izzie, so she strikes out in the third and walks in the fourth).

12. Butch has 1 rbi. That can't be in the first, third, fifth, or sixth. It must be in the second, so he must walk in the second and Chulo must have the remaining strikeout of the second.

13. The 6 runs of the second are the first 6 who walk, so Butch, Emma, and Fei Fei must score in the second. Since Dusty, Em, and Butch (second PA of the inning) strike out in the third, Butch (first PA of the inning) and Chulo walk. As the first 2 walks of an inning with 4 runs, they must score.

14. In the fifth inning Butch can't walk, or he would have to score the second run of the fifth, when we know he already had his runs in the second and third. So he struck out. Since Fei Fei and Hank also struck out, Chulo had to walk and score the second run of the inning. Since Dusty has no runs, Chulo must walk in the fourth to score the sole run of the inning.

15. Butch now has his 2 at-bats and so walks in the first and sixth. Chulo must strike out in the first and sixth to have his 3 at-bats.

Puppies	1	2	2	2	3	3	4	5	6	ab	R	H	rbi
Aiko	w		w	K		w		w	w	1	4	0	1
Butch	w		w		w	K		K	w	2	2	0	1
Chulo	K		K			w		w	w	K	3	3	0	0
Dusty	w		K		K		w	w	w	2	0	0	0
Em	w		w		K		w	w	w	1	1	0	4
Fei Fei	K		w		w		K	K	K	4	2	0	1
Gordito	K		w		w		w	w		1	1	0	4
Hank	w	w		w		K	K		2	1	0	2
Izzie	w	w		w		K		K	2	1	0	2
TOTALS	R 1 H 0		6 0		4 0		1 0	2 0	1 0	18	15	0	15

Solution to Macy at the Bat (p. 119)

In a game like this, a leadoff batter is always a leadoff batter. That means that there are three types of team play:

Type A. A team could have all nine players bat in each inning.

Type B. A team could bat all nine players over 2 innings. That could be 3 PA in the first followed by 6 PA in the second, with the pattern repeated two more times—3 6 3 6 3 6. Or it could be first 4 PA then 5 PA—4 5 4 5 4 5. Or the team could follow the pattern 5 4 5 4 5 4, or the pattern 6 3 6 3 6 3.

Type C. A team could have exactly 3 batters bat in each inning.

If Macy's team played a type A game, it would score 6 times whatever it scored in the first inning (and with 9 batters in the first it must score at least 3 runs). That didn't happen. If the game was type C then Macy's team would never score because there would always be just 3 batters per inning. So the game must have been of type B. But in each case, if Macy's team plays 6 innings, it should score three times whatever it scores in the first 2 innings.

But Macy's team scored only 2 runs. That's possible if it was the home team and was ahead after $5\frac{1}{2}$ innings and so didn't play the sixth. The only way that could happen is with the line score 0 1 0 1 0 x. Therefore Macy's team won. The score was 2-0, since if the other team scores as much as 1 run in 6 innings, it must score at least 3, and that would have forced Macy's team to play the sixth inning and then it would then have 3 runs too.

Note that a score of 2-0 couldn't occur in an extra-inning game, since if a team doesn't score in 6 innings, it will never score.

Solution to It's an ERG (p. 120)

Clearly Taipei didn't play exactly 6 innings, since (see the previous solution) 8 is not a multiple of 3. As in the first ERG game, Taipei could have played 5 innings but this time with a line score of 2 1 2 1 2 x. And of course, playing only 5 innings means Taipei won, as the home team.

There is another way Taipei can score 8 runs. The game could be tied after 6 innings, 6-6. Taipei could then win in 7 innings with the line score 2 0 2 0 2 0 2, while its opponent loses with 1 1 1 1 1 1 1 or with 0 2 0 2 0 2 0. Either way, Taipei wins.

Solution to Zhang and Li (p. 121); (Hint on p. 178)

First, the Tigers can't play a type A game. That's because Li, batting second, drives in all their runs, so they can score at most 2 runs per inning, but 9 batters per inning requires at least 3 runs per inning. On the other hand, Zhang, batting seventh for the Dragons, could drive in 3 or 4 runs per inning, so the Dragons, considered alone, could play a type A game. However, in that case the Dragons would score at least 3 runs per inning, while the Tigers, playing a type B or C game, could score at most 3 runs in each pair of innings. But then there is no way that the Dragons could win by just 1 run, which is required by the conditions of the problem. So the Dragons can't play a type A game either.

Next, neither team can play a game of type C (three players up every inning), because a type C team can't score any runs. We can't have a 0-0 tie. And even if only one team was of type C, it still wouldn't work, because again the winning team has to win by just one run, so the score would have to be 1-0. But scoring just 1 run in an ERG is impossible. So the games for both sides were of type B, with each team going through its lineup once every 2 innings. And again, since we can't have a 1-0 game, neither team was scoreless.

Playing a type B game, it is still true that Li will always bat second in every inning in which he bats, so the Tigers can score at most 2

runs in those innings. Thus, the runs for the Tigers repeat in patterns
of 2020 ... or 1010. With Zhang in slot 7, if the Dragons went out
1-2-3 in the first inning, he could drive in 3 runs in the second (but no
more, since there are at most 6 batters in the inning, of which 3 must
be outs). So the runs for the Dragons could be in the pattern of 0303
..., 0202 ..., or 0101 ....

Now suppose one of the games they played was a full 6 innings
with both teams getting 3 outs in the final inning. Then each team's
score would be a multiple of 3. But that's impossible, because we are
told it's a one-run game. For example, if the losing team scored 3 runs,
the winning team couldn't win 4-3, because 4 is not a multiple of 3.

We are left with two possibilities, a game ending in the sixth inning
but less than 6 full innings—call that a "6-minus game"—or a game
with more than 6 innings—call that a "6-plus game." In a 6-minus
game, the home team is the winner, either ahead after 5 1/2 innings, or
winning in a walk-off. For this reason, at most one of the games is a
6-minus.

In a 6-plus game, the winning team must score in the seventh. That
has to be the Tigers, since the Dragons score only in even innings. For
this reason, at most one of the games is a 6-plus. Thus we want one
6-minus game that the Dragons win and a 6-plus game that the Tigers
win. So the Dragons are the home team and the park is Taoyuan.
First we want a 6-plus game that the Tigers win as visitor. Since the
Dragons will score either 3, 6, or 9 runs, the Tigers must score 4, 7, or
10 runs. They could score 4 runs:

  1 0 1 0 1 0 1

but that's all. 2 0 2 0 2 0 2 gives 8 runs (and 3 0 3 0 3 0 3 can't be
done with Li batting second). The game is 7 full innings:

| Tigers: | 1 | 0 | 1 | 0 | 1 | 0 | 1 | —4 |
| Dragons: | 0 | 1 | 0 | 1 | 0 | 1 | 0 | —3 |

Then we must find a way for the Dragons to beat the Tigers in a
6-minus game by a score of 4-3. And here it is:

| Tigers: | 1 | 0 | 1 | 0 | 1 | 0 | —3 |
| Dragons: | 0 | 2 | 0 | 2 | 0 | x | —4 |

It lasts exactly 5 1/2 innings.

Third Bonus Puzzle!

The Tigers and the Dragons had a rematch! This time it was a "home and away" double-header, one game in one team's park and the second game in the other team's park. Again, they split the double-header, each team winning one game. Again, each game was decided by a single run, and the scores were identical—but not the same as in their previous double-header. How did this happen?

Solution on p. 257

Solution to Li and Zhang (p. 121)

Note first that Li, who has all of the Tigers' rbis, bats second, so the Tigers only score in odd innings. Zhang, on the other hand, bats seventh, so the Dragons only score in even innings. Among other things, that means that the Dragons can't score in a seventh inning and so can't win in extra innings.

As in the solution to Zhang and Li, the losing team must score 0, 3, or 6.

We have to consider each possibility.

If a losing team scored 0, then it would have to win the other game with a score of 8. A score of 8 is only possible in extra innings or in a walk-off, since neither team can score every inning. As in Zhang and Li, the Dragons can't do that. The only possibility is that the Tigers win in the seventh as either the visitor or the home team with a score of 8-6:

Tigers:	2	0	2	0	2	0	2
Dragons:	0	2	0	2	0	2	0

The other game would have to be 0-2 (for both Li and Zhang to drive in 8 runs). That's possible. The Dragons can win as the home team after 5 1/2 innings:

Tigers:	0	0	0	0	0	0
Dragons:	0	1	0	1	0	x

The Dragons are the home team and lose 8-6 then win 0-2.

Now for the second case, suppose the loser scores 3. Then it would have to win the other game scoring 5. But it's not possible to score 5. Recall each team can only score every other inning. So to get up to 5 runs by the sixth or seventh inning, you must score 2 runs per scoring inning, and then end with a one-run walk-off inning, which requires 0 2 0 2 0 1. But this is impossible because the losing team has only 3 runs after their sixth inning, so the home team wouldn't bat at all in the bottom of the sixth.

Finally, in the third case, losing with 6 runs, you can't lose 7-6, because that would require winning the other game 2-1, and it's impossible to score just 1 run in an ERG. So the losing team loses 8-6 and wins the other game 2-0. We already proved that only the Dragons can win 2-0, and thus the Tigers win the other game 8-6. So the result is the same as our first case, and the solution is complete.

It must be then that the scores of the two games are 8-6 and 2-0. The Dragons must be the home team in the second game, hence the home team in both.

Solution to Extreme ERGS (p. 122)

In 6 innings, a type A team has 54 plate appearances. A type B team has 27 PA. A type A team has 18. Clearly, the longest game must be between two type A teams. An ERG between two type As can't have extra innings—if it's tied after six, it will be tied after seven, etc. So the longest game has 108 PA.

Type B teams score at most 3 runs in 2 innings, while type As score at least six, so no extra-inning games between a type A and a type B are possible. Type C teams can't score at all so the only possible extra-inning games are between two type Bs.

After 6 innings, type B teams have batted around the order 3 times each, for 27 PA. So the two teams combined have 54 PA in the first 6 innings. How many more could they have in the seventh inning? Since type B teams have 3–6 PA per inning, could they both have 6 PA in the seventh inning? No, because that would require them both to have 6 PA in all odd innings and 3 PA in all even innings. That would require them to have 0 runs in all even innings, and score equally in odd innings, so the game would be judged infinite after 1 inning.

But we could have one team with 6 PA in the seventh inning, and the other with 5 PA, making a total of 54 + 11 = 65 PA for both teams. Here's one way: the visiting team sends six to bat in odd innings and scores in 7 innings: 2 0 2 0 2 0 2—8. The home team sends five to bat in odd innings and 4 in even innings and scores in 7 innings: 1 1 1 1 1 1 1—7.

The shortest possible game is infinite, but ends as soon as it is provably infinite. Here's one short game:

Both teams bring six players to the plate in the first, scoring equally. Then the coaches know that in the second inning each team will send only three players to the plate. That means they can't score in the second inning. So after just one inning, they know the game will be infinite and play ceases. The total PA is 12.

Another way to achieve this is a game in which both teams send three to the plate in the first and second innings. This establishes them both as type C teams. Again, the total PA is 12.

Solution to <u>Advanced ERGonomics</u> (p. 123); (Hint on p. 178)

Recall that there are three types of ERG patterns:

1. Type A: 9 PA per inning
2. Type B: 9 PA per 2 innings (single innings can have 3 to 6 PA)
3. Type C: 9 PA per 3 innings (3 PA per inning)

But the final inning can have fewer PA than what's listed above if it's by a team that wins in a walk-off or doesn't need its final half inning to win.

A team can have at most one more inning than the other team.

A game can have one of 6 combinations of the type A, type B, and type C teams.

Type C vs Type C has no runs, can never end, so isn't played out.

Type A vs Type B has a difference of 27 PA after 6 innings, and 1 inning more or less can only reduce the difference to 18. This doesn't work.

Type A vs Type C has a difference of 42 after 6 innings; 1 inning more or less can't reduce that to 8 PA.

Type B vs Type B has no difference after 6 innings and giving one team an inning more or less can only make a difference of at most 6.

Type B vs Type C: The type C team never scores, so for the game to end at all, the type B team must score and win. If the type B team is the visiting team, the game goes 6 full innings and it has 9 more PA (27 vs 18). If it's the home team, it wins after 5 1/2 innings and has 3 to 6 fewer PA, reducing the difference in PA to 3 to 6; not 8.

By elimination, the teams must both be of type A: 9 PA per inning.

The loser, a type A team, must play 6 whole innings with 54 PA. The winner, also a type A team, can't do the same or it would have the same number of PA. It must win in just 5 innings or in a sixth-inning walk-off. The winner, then, will have fewer PA than the loser. To satisfy the conditions of the puzzle, the winner should have 8 fewer PA, that is, 46 PA. But the winner will have 45 PA from the first 5 innings, so it will have to win with just 1 PA in the sixth.

A walk-off requires that the teams are tied after 5 1/2 innings. The loser's score after 6 innings is a multiple of six. The winner's score after 5 innings is a multiple of five. The tie score must therefore be $5 \times 6 = 30$, with the loser scoring 5 runs per inning for 6 innings and the winner scoring 6 runs per inning for 5 innings, followed by a single run in the sixth with 1 plate appearance (for example, with a solo home run). The final score is 31 to 30. The line score will look like:

5 5 5 5 5 5—30, 54 PA
6 6 6 6 6 1—31, 46 PA,
the sixth inning featuring a lead-off home run.

Solution to <u>Third Bonus Puzzle</u> (p. 253)

Recall that the home team wins 6-minus games and the Tigers win 6-plus games. The Dragons must win a game, so that's a 6-minus game. They win it in Taoyuan.

Once again, we need one 6-minus game (so the Dragons can win one) and one 6-plus game (so the Tigers can win one) The 6-plus game must be in Taoyuan so the Dragons can score. The scores have to be different, so instead of winning 4-3, the Dragons must win 7-6. That's the only possibility, since the Tigers will either have 3 runs or 6. But the Dragons can do it. It's a walk-off in the sixth:

Tigers:	2	0	2	0	2	0	—6
Dragons:	0	3	0	3	0	1	—7

This is possible. The Dragons could have 3 straight outs in the first, and in the second inning the #4, 5,and 6 batters could load the bases and all be driven in by Zhang on a triple. Then in the sixth inning Zhang's hit could be identical, but only one run would count for the walk-off win.

Could the Tigers win a 6-minus game by a score of 7-6? They can't. The most they can score in 6 innings is 6 (with Li batting in the 2 slot).

So we have to find a way for the Tigers to win a 6-plus game by a score of 7-6. To reach 7, the Tigers' first 6 innings must go: 2 0 2 0 2 0. Li bats second in the batting order and he drives in all the runs in the first, third, and fifth innings. The only way that can happen is if the #1 batter gets on base and then Li hits a home run. Good. But (Oh no!) in the seventh, when Li hits a home run, both runs count in the walk-off because the runs happen simultaneously[5] (unlike Zhang's triple in the other game).

So is this impossible? No, there's a way. The game works if Li's homer in the first inning (and therefore in the third inning) is an inside-the-park homer! That's when the ball stays in the ballpark instead of clearing the fences, but the batter still makes it home before

---

[5]They happen the instant the ball goes over the fence.

the fielders can put him out. Then, in the sixth inning, as soon as the batter in slot 1 scored, the game would be over. By our definition, this still counts as an ERG—the game eerily repeats until it is over. Li doesn't have to score, and can't score, because the game is over. And the score is 7-6.

Dragons:	0	2	0	2	0	2	0	—6
Tigers:		2	0	2	0	2	0	1	—7

Crazy Things That Don't Happen in This Book

In Chapter 2 we told you about bizarre events that can happen in baseball, events whose existence limits what we can deduce about games. We're going to tell you now about events that *don't* limit our deductions, because they will never come up in our puzzles.

I. CI and FO

There are two ways to get on base with no at-bat that weren't mentioned in the table in Chapter 1, p.18. The two ways are catcher interference (CI)—when the catcher interferes with the batter's swing—and fielder obstruction (FO)—when a fielder interferes with the batter running to first base.[1] If an umpire rules that one of these has occurred, the batter is granted first base and the catcher or fielder is charged with an error.

For afficionados, CI and FO affect the equation we have linking plate appearances and at-bats, because while they place the batter on first base, they are not counted as at-bats. The result is that the complete equation linking PA and ab is:

$$PA = ab + BB + HBP + S + SF + CI + FO,$$

—where by CI and FO, we mean the total numbers of times during the game that players are awarded first base because of catcher interference and fielder obstruction.

---

[1]Fielder obstruction can occur more broadly, when a runner is anywhere on the base paths. But here we are only concerned with fielder obstruction that affects a batter trying to reach first base.

In the equation mentioned in the previous page, we compute PA using our usual R + O + LOB, which is still true. A typical box score doesn't say anything about CI or FO. But if the equation balances without these terms, it proves that CI + FO is zero, i.e, that these events never occurred. If it doesn't balance without these terms, with the right hand falling short, then the difference is the number of CI and FO.

We have used the above method to check all the real box scores we use in this book, and none of them have any CI or FO. But if you don't trust us, you can check for yourself.

In two puzzles involving real games, Just Totals and the associated Bonus Puzzle, we don't provide totals for outs in the game, so it's not possible to check for CI and FO. But you may assume that are both zero, as they were in the actual games.

In all our composed puzzles, CI and FO are simply irrelevant because we never use the PA/AB equation, the only way it might affect them.

II. Batting with no one on base

In many places in this book, we argue that a batter who comes to the plate with the bases empty must have made an out because we've ruled out all the ways he might have gotten on base: hits, walks, HBP, uncaught third strikes, or errors (which includes errors that are assessed for CI or FO). However, one can imagine some ways to get on base with the bases empty without any being scored a hit or involving an error. The simplest ones involving an infielder who fielded a ground ball in time to make an out but never threw to first base.[2] We (the authors) simply rule them out. We have checked the play-by-play records of all the real games used in the book to assure that they don't occur in these games. In any puzzle in this book, you may take it as a premise that such events don't happen.

---

[2]Why wouldn't he throw to first? Perhaps the infielder thought he caught a low line drive, but the umpire thought he trapped it and ruled the batter safe. Perhaps the infielder was daydreaming, thought there was a baserunner, and threw to the wrong base. Or maybe he just wanted to let the batter get on base to help him set a stolen base record in a meaningless game. We haven't been able to document any instances of these things actually happening.

III. Batting out of turn.

The box score lists the batters in the order they are supposed to bat. But very occasionally they accidentally bat out of turn. That can mess up our deductions.

Suppose a batter simply skipped his turn and nobody noticed. The box score would show the correct lineup, but our usual PA method would identify the wrong final batter in the game. All our deductions working backward would be wrong.

Suppose a batter simply skipped his turn. Then he wouldn't score a run, he wouldn't be left on base, and he wouldn't be put out. That would make our PA equation faulty!

If we allowed for batting out of order, we'd have to scrap most of this book. But there's an easy fix—we're informing you that none of the box scores we have chosen to use are for games in which anybody batted out of order. And we hereby confirm our tacit assumption that nobody bats out of order in our composed problems either.

For those reasons, in any problem we compose, you may assume that nobody bats out of order. And whenever analyzing a real box score, we will (and you may too) assume that no one has batted out of turn.

We (the authors) have worried about this so that you (the readers) don't have to worry.[3]

IV. Ghost runners

When the COVID-19 pandemic hit, major league games played under several new rules. One is the "ghost runner" rule. In extra innings, every half-inning begins with a player from the team at bat being placed on second base. This is the "ghost runner." The player chosen as ghost runner in each half inning is the player who made the final out for the team in the previous inning.

---

[3]It is a fact, however, that every other year or so, some major leaguer bats out of turn. But don't worry. We're on the job.

While the ghost runner can be driven in and score a run, the ghost runner is not given an at-bat. His appearance, however, has the effect of a successful plate appearance. That messes up a few of our equations. For this reason, readers may assume that all ball games are played under pre-pandemic rules, i.e., no ghost runner. There is a brief exception to this in Chapter 7. Readers are clearly informed.

While some pandemic rules have been allowed to lapse, the ghost runner rule is still in place, as of this writing.

Printed in the United States
by Baker & Taylor Publisher Services